# Improving Acquisition to Support the Space Enterprise Vision

YOOL KIM, GUY WEICHENBERG, FRANK CAMM,
BRIAN DOUGHERTY, THOMAS C. WHITMORE,
NICHOLAS MARTIN, BADREDDINE AHTCHI

Prepared for the United States Air Force
Approved for public release; distribution unlimited

For more information on this publication, visit www.rand.org/t/RR2626

**Library of Congress Cataloging-in-Publication Data** is available for this publication.

ISBN: 978-1-9774-0539-5

Published by the RAND Corporation, Santa Monica, Calif.

# Preface

The national security space community recognizes that the threat to the space domain is increasing and, as a result, the current U.S. space enterprise is not resilient enough to survive a conflict that extends to space.[1] At the time this report was written, Air Force Space Command (AFSPC) was leading an effort to build a space force that will prevail in the face of the threat. Specifically, AFSPC envisioned an integrated approach—the Space Enterprise Vision (SEV)—to building a force across all space mission areas, coupling the delivery of space capabilities to the warfighter with the ability to protect and defend space capabilities against emerging threats. Since that time, the U.S. Space Force has stood up as a separate service under the Department of the Air Force, and the implementation of the SEV has transitioned to the Space Force.

The SEV requires significant changes in how the Air Force designs, develops, acquires, and operates its space systems to build a future space enterprise. With respect to the acquisition of space systems, the SEV demands more responsive acquisition of systems and improved synchronization of systems across the enterprise. AFSPC and the Space and Missile Systems Center (SMC) asked RAND Project AIR FORCE (PAF) to help the Air Force identify ways to achieve those SEV acquisition goals. This document presents RAND PAF's investigation into how space acquisition could be improved to support the SEV. In particular, this research focused on three promising acquisition approaches to achieving the SEV goals: modular open systems

[1] AFSPC Public Affairs, "AFSPC Commander Announces Space Enterprise Vision," April 11, 2016.

architecture (MOSA), agile acquisition, and rapid prototyping. The challenges and recommendations identified in this report remain relevant to the Space Force as the new service aims to implement an alternative acquisition system to accelerate space acquisition and improve synchronization across space, ground, and user segments.

This research was sponsored by the Director of Strategic Plans, Programs, Requirements and Analysis, Headquarters Air Force Space Command, and the Vice Commander, Space and Missile Systems Center, and was performed within the Resource Management Program of RAND Project AIR FORCE. The research was conducted between October 2016 and August 2017. This document should interest those involved in the planning and implementation of the SEV and those engaged in the acquisition of space systems.

## RAND Project AIR FORCE

RAND Project AIR FORCE (PAF), a division of the RAND Corporation, is the U.S. Air Force's federally funded research and development center for studies and analyses. PAF provides the Air Force with independent analyses of policy alternatives affecting the development, employment, combat readiness, and support of current and future air, space, and cyber forces. Research is conducted in four programs: Strategy and Doctrine; Force Modernization and Employment; Manpower, Personnel, and Training; and Resource Management. The research reported here was prepared under contract FA7014-16-D-1000.

Additional information about PAF is available on our website: http://www.rand.org/paf/

# Contents

# Figures and Tables

## Figures



## Tables

# Summary

For many years, the United States has had unimpeded access to space, enabling delivery of a multitude of capabilities—intelligence collection, missile warning, weather monitoring, satellite communications, and precise positioning, navigation, and timing—to U.S. armed forces operating globally. Consequently, the U.S. space enterprise was conceived with long-term functionality and efficiency in mind, rather than resiliency to potential attacks on space capabilities. However, space is no longer a sanctuary within which the United States may operate with impunity, because adversaries are developing various means to disrupt U.S. space capabilities. Future space systems must therefore be developed, produced, and deployed with this contested space environment in mind.

The Space Enterprise Vision (SEV), developed jointly by Air Force Space Command (AFSPC) and the National Reconnaissance Office in 2015, describes an integrated approach to building a force across all space mission areas, coupling the delivery of space capabilities to the warfighter with the ability to defend space capabilities against emerging threats. Achieving this vision requires reducing acquisition time lines and improving integration of the space enterprise. Given that the SEV requires a departure from the way space systems are currently acquired, in FY 2017, AFSPC headquarters and Space and Missile Systems Center (SMC) asked RAND Project AIR FORCE (PAF) to assess key barriers to realizing the SEV—to include requirements, acquisition, and funding processes, as well as industrial base considerations—and recommend ways to overcome those barriers to help achieve the SEV goals.

The research team examined a range of potential approaches to support the goals of the SEV, based on a literature review and semistructured interviews with acquisition subject matter experts and sponsor guidance. PAF identified several promising alternative acquisition approaches that merit in-depth examination in this project:

- modular open systems architecture (MOSA)
- agile acquisition
- rapid prototyping.

These concepts are not new, but implementing them in a Department of Defense (DoD) setting is challenging for a variety of reasons. Therefore, we focused our examination on likely challenges to implementation and made recommendations to overcome them to improve the likelihood of success should these acquisition approaches have a role in the SEV.[1] Each of these acquisition approaches can be implemented independently, but there are instances in which they may be used synergistically in conjunction with each other, as we illustrate in the body of the report with the case of the Enterprise Space Battle Management Command and Control (ESBMC2) program.

## Modular Open Systems Architecture

A MOSA system is characterized by a modular design with loosely coupled, plug-and-play system modules that are connected via widely supported, consensus-based interface standards. Implementing MOSAs in the SEV offers the potential to upgrade capabilities more responsively and enable enterprise synchronization in such areas as a common ground system or common satellite bus. While the concept of MOSA is relatively straightforward, its implementation within the DoD acquisition-

---

[1] There are many barriers and challenges to the full and successful application of these approaches, and they may vary depending on the specific program. This research does not fully examine the applicability of these approaches to a range of space acquisition programs that are in the Air Force's portfolio; nor does it provide a comprehensive risk analysis to assess potential risks of employing or adopting these approaches.

management framework requires a much deeper understanding and appreciation of the complexities associated with it. Successful implementation of MOSA in a DoD environment involves overcoming several challenges.

- *Aligning diverse stakeholders when MOSA is applied to a large enterprise.* Common interface standards for a broad enterprise enable interoperability and technology reuse across the enterprise, but their acceptance and adoption require agreement of diverse stakeholders on common goals and requirements—which in some cases may result in suboptimal program-level decisions for the greater good of the enterprise or portfolio. Obtaining consensus in this context can be challenging because individual program managers are evaluated on the success of their program, not enterprise performance. Such concepts as common ground system or common satellite bus would face this challenge because of the stovepipe nature of the current space and ground architectures.
- *Managing acquisition through incremental development and fielding.* MOSAs are designed to take advantage of flexible, modular architectures to quickly and incrementally develop capability improvements, which is counter to the rigidity of legacy requirements and acquisition processes. Although more recent acquisition governance documents direct tailoring of acquisition processes, taking advantage of the dispersed DoD expertise on how to design and implement incremental programs is a challenge to successful implementation of MOSA. Incremental development and fielding of the space segment may be more challenging than the ground segment because of the difficulties in upgrading on-orbit hardware. This issue is further compounded by long satellite lifetime that limits opportunities for frequent upgrades.
- *Determining key interfaces that should be open.* Not every interface involved in connecting system elements must be standardized. MOSA system owners must determine which key interfaces should be standardized and governed by an interface standard and which do not need to be. Developing standards and ensuring their success involves significant long-term management and support and is a complex process involving many stakeholders.

- *Tailoring engagement with industry on standards and data rights.* The low-volume space market has hampered voluntary standardization within the space community. In the absence of a driving force with large market share, achieving common standards in space could continue to be challenging. Past initiatives for developing common standards for satellites have failed due to the low-volume issue and lack of government support. Additionally, some level of data rights may be needed for the integration of MOSA's subcomponents or for recompeting individual subcomponent contracts.[2]
- *Selecting an appropriate system integration model.* In a DoD setting, two potential integration models exist for MOSA, depending on which organization will be responsible for system integration: either the government or a contractor. Using government or contractor integration each has its own challenges and benefits. Ensuring the success of MOSA involves the long-term evaluation of these trade-offs to determine the appropriate integration model for a given development.

**Recommendations**

To overcome the identified challenges and successfully use MOSAs to achieve SEV goals, the Air Force should enact policies that enable MOSAs to take hold to the maximum extent practical. Our recommendations and associated actions—distilled from lessons and best

---

[2] Although data rights might not be central to MOSA *in theory*, they have shown to be an important factor in realizing the benefits of MOSA in past programs (Acoustic Rapid COTS Insertion [A-RCI] Program Office staff, discussion on A-RCI lessons on MOSA implementation, Washington, D.C., January 31, 2017; Boeing staff, discussion on MOSA, agile, and rapid prototyping, El Segundo, Calif., June 29, 2017; Air Force Office of Transformational Innovation [OTI] staff, discussion with authors on Other Transaction Authority [OTA], Open Architecture, Arlington, Va., May 31, 2017b; Frank Camm, *Analysis of Technology Innovator Industrial Base for Joint Counter Radio-Controlled Improvised Explosive Device Electronic Warfare 3.3*, Santa Monica, Calif.: RAND Corporation, 2011, Not available to the general public). Reopening of competition could be desirable for sustainment reasons, to encourage innovation and cost reduction, or to groom alternate suppliers in the event the primary supplier would leave the market. In these cases, to reopen production competition for a component already in production, a design agent must own sufficient data rights to allow another supplier to produce the given component.

practices from successful programs, discussions with subject-matter experts, and guidebooks on MOSA applications—are summarized in Table S.1, which indicates whether they apply to AFSPC, SMC, or both.

*Establish an enterprise governance.* An environment with an established enterprise, shared goals, and enterprise-level governance helps decisionmakers make trade-offs among the needs of diverse stakeholders. To successfully implement an enterprise governance structure, AFSPC and SMC should ensure the new governance structure has: (1) goals and outcomes that are tied to the Air Force space enterprise's vision/goals, (2) members with sufficient data and/or knowledge to make informed decisions, (3) sponsorship by senior leadership, and (4) formal structure with defined and repeatable processes and performance metrics to measure success.

**Table S.1**
**Recommendations for MOSA**

| Recommendation | AFSPC | SMC |
|---|---|---|
| *Overarching Recommendation 1: Establish an Enterprise Governance* | | |
| Define the enterprise scope and relationship to organizational goals and incentives | X | X |
| Provide governance members the data needed to inform decisionmaking | X | X |
| Obtain executive support and formalize metrics-based evaluation processes | X | X |
| *Overarching Recommendation 2: Invest in Up-Front Systems Engineering and Infrastructure* | | |
| Develop a functionally partitioned design | | X |
| Select appropriate interface standards | | X |
| *Overarching Recommendation 3: Develop Up-Front Life-Cycle Business and Program Plans* | | |
| Develop business engagement plans to enable competition | | X |
| Develop integrated life-cycle program plans | | X |

NOTE: An "X" indicates relevance to the organization in the corresponding column.

*Invest in up-front systems engineering and infrastructure.* Once an enterprise governance structure is established, AFSPC and SMC can begin to consider the technical aspects of implementing a MOSA. This includes development of a technical plan to create functionally partitioned design connected by broadly adopted standards and/or middleware. Without a broad base of users, even an openly developed standard provides no additional technical or business benefit to a proprietary standard, and support for such a standard will wane as its user base decreases.

*Develop up-front life-cycle business and program plans.* In parallel with development of a technical plan, AFSPC and SMC should consider program business engagement and programmatic plans. Business engagement plans include data rights strategies and selection of an integrator. Programmatic plans involve collaborative interaction on mandatory program structure questions and early development of a program life-cycle plan. This includes using operational considerations to inform development, testing, fielding, and sustainment time lines. Successful MOSA programs, such as the Navy's A-RCI program, may serve as a useful model.

## Agile Acquisition

Agile acquisition is evolved from quality management methods now associated with Lean Six Sigma—methods that produce things "better, faster, and cheaper"—to improve the development and maintenance of software.[3] Today, agile acquisition is most commonly applied to software development, but its use in hardware development is growing. Agile acquisition is characterized by incremental development and the ability to react quickly to changes in capability needs.

Federal standards associated with requirements determination, acquisition rules, and budgeting systems often constrain the applica-

[3] Agile acquisition shares attributes with earlier approaches to evolutionary and spiral development but has moved well beyond them in the processes, methods, tools, and skills that it applies.

tion of agile acquisition practices within a government setting. But the number of success stories is growing, which offers insights into how agile practices can be accommodated and highlight implementation challenges that repeatedly arise. Those most relevant to SEV are as follows:

- *Tailoring agile acquisition to specific circumstances and choosing the right hybrid.* Agile acquisition practices are not appropriate for every acquisition setting. Depending on the setting, agile, traditional, or a hybrid may be the best approach, and it can be difficult to determine which approach is best. Getting the right mix of traditional compliance and agile tailoring requires an understanding of how applicable different federal regulations are to the agile process and how they achieve desired outcomes.
- *Defining requirements to get full benefit of agile acquisition methods.* Traditional acquisition approaches formulate specific capability requirements, which are translated into engineering specifications in a strict linear approach. With agile, the approach to requirements determinations is to emphasize capability rather than engineering specification.
- *Ensuring that developers and those who oversee them have appropriate training and experience.* Government personnel regularly face challenges implementing agile approaches when they rely on developers that lack relevant experience and when they themselves are not experienced enough to get the most from their contractors. Appropriate training can ensure that government personnel are familiar with agile approaches.[4] Training applies not only to the defense acquisition workforce but also needs to extend to requirements developers and operational users, among others that enable or support acquisition. The Air Force may need to look to nontraditional sources of training since little training on these concepts is available from government sources, such as the Defense Acquisition University.

---

[4] DoD personnel may feel they are not incentivized to take risks associated with using a new approach that they are not familiar with.

- *Maintaining timely, committed user participation.* Users—or representatives of their interests as stakeholders—participate much more in an agile acquisition than they would if the system was acquired in a traditional process. One of the challenges in obtaining user participation is that government organizations must commit the time of these users, typically by removing other commitments. Additionally, there needs to be continuity in the users engaged in development over a series of increments. Also, the government must choose which user or user representatives to involve in development activities; this can be particularly challenging for space acquisition because space-based systems often provide services to diverse users well removed from the Air Force.

**Recommendations**

Our analysis identified potential ways to overcome these challenges to improve the likelihood of successful implementation of agile acquisition in the SEV. Table S.2 details our three overarching recommendations and associated actions.

*Build and sustain an experienced workforce for agile acquisition.* Agile acquisition works best when the people involved have sufficient experience and skills to make appropriate decisions without detailed instructions. These qualifications apply to developers, those who work with them, and those who oversee all elements of the acquisition from requirements determination to software sustainment. AFSPC and SMC will likely need a formal workforce-shaping program to achieve the appropriate expertise to implement agile acquisition where it is appropriate to use. This will involve building a workforce with appropriate skills and continuity working together, which takes time but can be accelerated through staff rotations at other organizations with experience or bringing in expertise from the outside. It will also require building the support of the senior Air Force leadership by bringing in specialists that can help obtain and sustain leadership engagement and support. Another policy consideration is to make agile acquisition experience a prerequisite to leading a major software development or acquisition effort. Active engagement with the external environment to stay abreast of best practices is especially helpful for agile acquisition.

**Table S.2**
**Recommendations for Agile Acquisition**

| Recommendation | AFSPC | SMC |
|---|---|---|
| *Overarching Recommendation 1: Build and Sustain an Experienced Workforce for Agile Acquisition* | | |
| Build a workforce with appropriate skills and continuity working together | X | X |
| Build support of the senior Air Force leadership | X | X |
| Make agile acquisition experience a prerequisite for advancement in areas where you expect to use agile acquisition in the future | X | X |
| Maintain regular contact with others conducting agile acquisition to stay abreast of best practices | X | X |
| *Overarching Recommendation 2: Create an Enabling Environment for Agile Acquisition* | | |
| State requirements to allow ongoing flexibility as relevant threats and technologies change | X | |
| Work with relevant user communities to ensure effective user participation | X | X |
| Work with the contracting community to create a seamless way for users to direct developer activities in compliance with Federal Acquisition Regulation (FAR) | | X |
| Work with the test community to certify and coordinate internal and external tests focused on realized operational capability | X | X |
| *Overarching Recommendation 3: Build "Agile" into Acquisition Strategy* | | |
| Frame the request for proposal (RFP) and contract materials to facilitate agile acquisition | | Mainly |
| Build a calendar-driven battle rhythm that assesses status at fixed intervals | | X |
| When possible, rely on trust, reputation to ensure delivery of what is promised to limit the need for active oversight | | Mainly |

NOTE: An "X" indicates relevance to the organization in the corresponding column. "Mainly" indicates relevance to both but primary relevance to SMC.

*Create an enabling environment for agile.* Early critics of agile acquisition thought it lacked the discipline of plan-driven acquisition. In fact, agile acquisition requires close discipline to create an environment that is safe for rapid learning and creativity. Advance planning is required to prepare such an environment and must involve the relevant user communities, requirements developers, the contractor community, and the test community. Agile acquisition takes the time to do this planning because it allows much more rapid development once development actually begins. Much of this planning involves ensuring that government and contractor development teams have the experience, skills, and tools to work well together with limited top-down oversight.

*Build "agile" into acquisition strategy.* An agile acquisition strategy starts with an RFP that frames the type of contract that SMC will use to access developers. Contracts should be flexible and should incentivize behavior likely to facilitate effective agile acquisition. Agile acquisition typically proceeds through sprints and releases that occur during fixed periods of calendar time, completing whatever work can be finished during a planned period. Developers tend to determine the sprints and releases, but SMC must understand what to expect, ensure that all developers are effectively integrated, and ensure there are common expectations about program timing. In addition, agile acquisition establishes a different relationship between developers and users that is based on a mutual trust developed over time.

## Rapid Prototyping

Prototyping is typically characterized as a set of design and development activities intended to reduce technical uncertainty and to generate information to improve the quality of subsequent acquisition decisionmaking. It is traditionally accomplished in the early stages of an acquisition program of record (POR) to reduce technical risk before proceeding into initial production and the operations and support phase of an acquisition program. Our research focused on a narrow application of rapid prototyping that AFSPC is pursuing: we characterize *rapid prototyping* as experimentation conducted outside the POR to field a capability faster than a traditional acquisition POR.

As AFSPC develops new space architectures and operational capability requirements to implement the SEV, we foresee three circumstances in which AFSPC may want to conduct a rapid prototyping activity: to respond rapidly to technological surprise by an adversary that poses a threat to space systems and capabilities; to implement planned enhancements to space architectures and systems; and to demonstrate and evaluate promising, innovative technologies in an operational setting.

In the course of our research, we discovered a number of challenges the Air Force often faces when attempting to transition prototype technologies or capabilities into operational service and into follow-on acquisition PORs to sustain and further improve the prototype capability.

- *Securing user support and funding source.* First among these challenges is the need to secure operational user support to accept the prototype capability and source the funding needed to operate, sustain, and mature the capability. This challenge is very different from the traditional problems associated with laborious defense acquisition processes that are chronically slow to deliver products to an operational community in a timely manner. This challenge is often complicated when operational users are skeptical of the prototype's operational effectiveness and unwilling to commit to unproven technologies.
- Additionally, acquisition and operational climates dominated by risk aversion and fear of failure (of developing a prototype that does not meet the user's critical needs) tend to reduce the likelihood of successful prototype transition. Pressure on the Air Force's budget, where new systems or capabilities must compete for funding against legacy capabilities and programs with a proven history of success, also limit the ability to transition innovative prototype systems into operational service.
- *Addressing operation and acquisition considerations after the fact.* Other challenges to prototype transition surface when operational

and acquisition considerations are not adequately addressed as the prototype is being developed and tested. When such factors as technical maturity; manufacturing robustness; product support and sustainment; and operational tactics, techniques, procedures, and training requirements are addressed *after* the prototype is designed, built, and tested, the likelihood of timely and successful transition to field use is also diminished.

**Recommendations**

Based on our literature review and interviews with government and defense industry subject-matter experts, we developed two overarching recommendations AFSPC should consider to overcome the challenges of operationalizing a rapid prototype that are shown, along with supporting actions, in Table S.3.

*Institutionalize a prototype development, evaluation, and transition process.* AFSPC should institutionalize a disciplined, repeatable, and measurable prototype development, evaluation, and transition process to provide the operational community with the details needed to make an informed decision on acceptance. This entails adopting and employing technology/prototype transition best practices, which are detailed in this report. AFSPC also needs to codify rapid prototyping expectations, events, and transition agreements, such as the Navy's Future Naval Capabilities (FNC) program processes and Technology Transition Agreement (TTA). In addition, AFSPC should consider adopting the U.S. Special Operations Command (USSOCOM) Transition Confidence Level (TCL) scale as a means to track transition readiness.

*Associate and integrate rapid prototyping with acquisition PORs as early as possible.* AFSPC should address and develop follow-on requirements documents, program/budget requirements, and acquisition program strategies and plans concurrent with prototyping activities. Further, AFSPC should integrate rapid prototyping activities with operational fielding and employment plans using AFSPC Instruction 10-605, *Operational Acceptance Process*, as a guide. Finally, AFSPC should leverage recent legislative initiatives related to prototyping contained in FY 2016 and FY 2017 NDAAs.

**Table S.3**
**Recommendations for Rapid Prototyping**

| Recommendation | AFSPC | SMC |
|---|---|---|
| *Overarching Recommendation 1: Institutionalize a Disciplined, Repeatable, and Measurable Prototype Development, Evaluation, and Transition Process* | | |
| Adopt and employ technology/prototype transition best practices | X | X |
| Codify governance, process, and agreements for prototype transition | X | X |
| Track transition readiness using USSOCOM Transition Confidence Level | X | X |
| *Overarching Recommendation 2: Associate and Integrate Independently Conducted Rapid Prototyping Activities with an Acquisition POR as Early as Possible* | | |
| Integrate rapid prototyping activities with operational plans | X | |
| Integrate rapid prototyping activities with acquisition POR content, ensuring Joint Capabilities Integration and Development System artifacts, Program Objective Memorandum (POM)/budget requirements, and acquisition program strategies and plans are addressed | X | X |
| Leverage recent National Defense Authorization Act (NDAA) provisions related to rapid prototyping | X | X |

NOTE: An "X" indicates relevance to the organization in the corresponding column.

## Posturing for Success

More responsive acquisition can improve the resilience of the space enterprise and help achieve the goal of the SEV. MOSA, agile acquisition, and rapid prototyping are three promising acquisition approaches that could be implemented as AFSPC evolves its space and ground architectures. These concepts require a different mind-set than traditional acquisition approaches. Thus, there are overarching policy-level actions that senior leadership in AFSPC and SMC may need to take to better posture the space enterprise for successful implementation.

- *Shift in architectures.* MOSA, agile acquisition, and rapid prototyping may have had limited opportunities for application in space acquisition because of certain structural issues in the space enterprise, such as low-volume space market, long spacecraft life, or stovepipe architecture (e.g., satellite-specific ground system). These issues need to be addressed in future architectures.
- *Investments in education and training of the workforce.* Space acquisition has had limited experience with MOSA, agile acquisition, and rapid prototyping. Effective implementation will require investments in human capital to build experience in the workforce and enable them to employ these concepts effectively.
- *Shift in organizational culture.* These three acquisition approaches emphasize different aspects of acquisition performance compared to traditional acquisition approaches—responsiveness over system performance and benefits for the enterprise versus an individual program—and, thus, will require a shift in organizational culture to look at risk and requirements in a different way.

Ultimately, AFSPC and SMC should determine the scope and applicability of these approaches in the SEV and consider the trade-offs associated with them. While responsiveness (and enterprise synchronization, in the case of MOSA) would be primary reasons for employing these approaches, there are other trade-offs that warrant further consideration as discussed in detail in the body of this report. Posturing for success entails big muscle movements: changes in the space enterprise architectures, investments in human capital, and shifting the organizational culture. Should MOSA, agile acquisition, and rapid prototyping have a role in the SEV, we suggest that formal change-management tools be employed to assist with implementation, needed collaboration, and the broader organizational changes that these alternative acquisition approaches call for.

# Acknowledgments

We thank Maj Gen Nina M. Armagno, Brig Gen Mark A. Baird, Maj Gen John E. Shaw, and Brig Gen Philip A. Garrant for sponsoring this work. We are also grateful for the support of our action officers, Maj Robert C. Small and Capt Shaun C. Cain, who opened many doors for us and provided helpful insights throughout the project. We also benefited from many discussions with individuals throughout the Air Force and industry, too numerous to mention by name.

At RAND, we thank Phil Anton, David Galvan, and Mark Lorell for their insightful comments, which further improved the report. We are also thankful to Obaid Younossi, Barbara Bicksler, and Jerry Sollinger, who provided helpful critiques and other support throughout this project.

# Abbreviations

| | |
|---|---|
| AEHF | Advanced Extremely High Frequency |
| AFRCO | Air Force Rapid Capabilities Office |
| AFRL | Air Force Research Laboratory |
| AFSPC | Air Force Space Command |
| AFSPCI | Air Force Space Command Instruction |
| AOC | Air Operations Center |
| A-RCI | Acoustic Rapid COTS Insertion |
| BMC2 | battle management command and control |
| C2 | command and control |
| CDD | Capability Development Document |
| CMCC | Common Mission Control Center |
| CO | contracting officer |
| COA | course of action |
| COTS | commercial off-the-shelf |
| CPAF | cost-plus-award-fee |
| CPD | Capability Production Document |
| DAU | Defense Acquisition University |

| | |
|---|---|
| DCGS | Distributed Common Ground System |
| DIUx | Defense Innovation Unit Experimental |
| DoD | Department of Defense |
| DoDD | DoD Directive |
| EGS | Enterprise Ground Service |
| EMD | Engineering and Manufacturing Development |
| ESBMC2 | Enterprise Space Battle Management Command and Control |
| EVM | earned-value management |
| FAR | Federal Acquisition Regulation |
| FNC | Future Naval Capabilities |
| FY | fiscal year |
| GAO | Government Accountability Office |
| GPS | Global Positioning System |
| GOTS | government off-the-shelf |
| ICD | Initial Capability Document |
| IT | information technology |
| JCIDS | Joint Capabilities Integration and Development System |
| JEON | Joint Emergent Operational Need |
| JFCC Space | Joint Functional Component Command for Space |
| JMS | Joint Space Operations Center Mission System |
| JSpOC | Joint Space Operations Center |
| JUON | joint urgent operational need |

| | |
|---|---|
| MMSOC | Multi-Mission Satellite Operations Center |
| MOSA | modular open systems architecture |
| NASA | National Aeronautics and Space Administration |
| NDAA | National Defense Authorization Act |
| NSDC | National Space Defense Center |
| O&S | operations and support |
| OMS | Open Mission Systems |
| ORS | Operationally Responsive Space |
| OTA | Other Transaction Authority |
| PAF | RAND Project AIR FORCE |
| PM | program manager |
| POM | Program Objective Memorandum |
| POR | program of record |
| R&D | research and development |
| RDT&E | research, development, test, and evaluation |
| RFP | request for proposal |
| S&T | science and technology |
| SAF/AQR | Assistant Secretary of the Air Force for Acquisition, Technology, and Logistics, Science, Technology, and Engineering Directorate |
| SBIRS | Space Based Infrared System |
| SEI | Software Engineering Institute |
| SEV | Space Enterprise Vision |
| SMC | Space and Missile Systems Center |
| SSA | space situational awareness |

| | |
|---|---|
| STRG | Submarine Tactical Requirements Group |
| SUMO | Space Universal Modular Architecture |
| TCL | Transition Confidence Level |
| TRL | Technology Readiness Level |
| TTA | Technology Transition Agreement |
| UCI | Unmanned Aerospace Systems Command and Control Initiative |
| UON | urgent operational need |
| USC | United States Code |
| USSOCOM | U.S. Special Operations Command |
| USSTRATCOM | U.S. Strategic Command |

CHAPTER ONE

# Introduction

> In the recent past, the United States enjoyed unchallenged freedom of action in the space domain. Most U.S. military space systems were not designed with threats in mind, and were built for long-term functionality and efficiency, with systems operating for decades in some cases. This is no longer an adequate methodology to equip space forces.
>
> —Gen John E. Hyten
> Former Commander, U.S. Air Force Space Command[1]

For many years, the United States has had unimpeded access to and operation in space, enabling delivery of a multitude of capabilities—intelligence collection, missile warning, weather monitoring, satellite communications, and precise positioning, navigation, and timing—to U.S. armed forces carrying out operations globally. Consequently, as expressed by General Hyten, former Commander of Air Force Space Command (AFSPC), the U.S. space enterprise was conceived with long-term functionality and efficiency in mind, rather than resiliency to potential attacks on space capabilities. However, space is no longer a sanctuary within which the United States may operate; adversaries are developing various means to disrupt U.S. space capabilities (Figure 1.1). Future space systems must therefore be developed with this contested space environment in mind.

---

1 AFSPC Public Affairs, "AFSPC Commander Announces Space Enterprise Vision," April 11, 2016.

**Figure 1.1**
**Possible Attack Vectors to Deny, Disrupt, Degrade, or Destroy Space Capabilities**

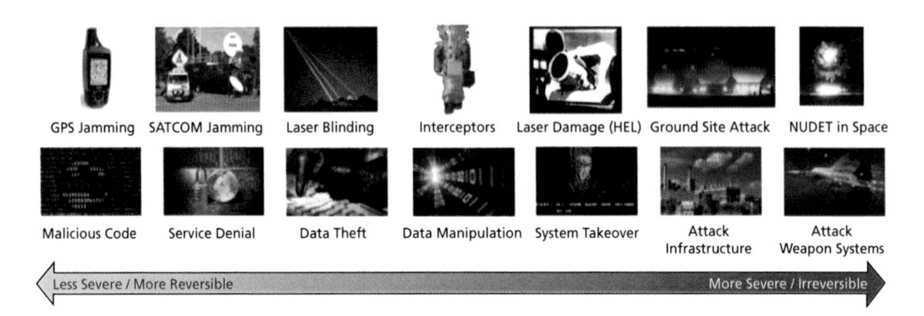


SOURCE: AFSPC, "Space Warfighting Construct," briefing slides, May 2017a.

The Space Enterprise Vision (SEV), developed jointly by AFSPC and the National Reconnaissance Office in 2015, describes an *integrated approach* to building a force across all space mission areas, coupling the delivery of space mission effects to the warfighter with the ability *to protect and defend space capabilities against emerging threats.*[2] It requires changes in how the Air Force designs, develops, acquires, and operates its space systems to build a future space enterprise. Several key attributes of the SEV have significant implications for the acquisition of space systems:[3]

- shorter program life cycles and decreased time between constellation updates to improve the availability of new technology on orbit
- mission assurance as the dominant enterprise requirement (rather than functional availability or reliability engineering)
- acquisition and programmatic decisions driven by an overarching space mission enterprise context, rather than in space mission stovepipes.

---

2 AFSPC Public Affairs, 2016.

3 AFSPC, "AFSPC-NRO Shared Strategic, Space Enterprise Vision," briefing, July 2016, Not available to the general public.

These attributes will require reducing acquisition time lines and improving integration of the space enterprise. Additionally, these attributes may have implications for how the space industrial base may need to evolve to support the SEV goals. Given that the SEV requires a departure from the way space systems are currently acquired, AFSPC headquarters and Space and Missile Systems Center (SMC) asked RAND Project AIR FORCE (PAF) to assess key barriers to realizing the SEV from an acquisition perspective—to include requirements, acquisition, and funding processes, as well as industrial base considerations—and recommend ways to overcome those barriers to help achieve the SEV goals.

## Research Approach and Scope

The research approach employed in this project comprised the following four tasks:

**Task 1: Identify promising alternative acquisition approaches to achieve responsiveness and enterprise synchronization.**
We conducted a preliminary literature review of past Department of Defense (DoD) programs (beyond space programs) and initiatives that have attempted to achieve responsive acquisition and enterprise synchronization to identify an initial set of candidate acquisition approaches to consider.[4] We also held discussions with acquisition offices focused on rapid acquisitions—Air Force Rapid Capabilities Office (AFRCO), the ORS Office, and the Air Force Office of Transformational Innovation (OTI)—to elicit information on ways to improve responsiveness in

---

[4] These programs included Acoustic Rapid COTS Insertion (A-RCI), Joint Space Operations Center Mission System (JMS) Increment 2, Joint Counter Radio-Controlled Improvised Explosive Device Electronic Warfare Program System Design, Predator, Space Based Space Surveillance, Space Universal Modular Architecture (SUMO), Multi-Mission Space Operations Center (MMSOC), Rapid Attack Identification, Detection, and Reporting System, Operationally Responsive Space (ORS) Office initiatives, Air Force Aeronautical Systems Division's Propulsion System Program Office, and F-16 multinational staged improvement program.

acquisition.[5] This initial review revealed several candidates as promising acquisition approaches. They included the use of open architectures, commonality, commercial off-the-shelf (COTS) products, commercial services and practices, evolutionary acquisition, agile acquisition, prototyping, and many elements and variations of these acquisition approaches.

Based on this initial review and guidance from our sponsor, we selected three acquisition approaches for in-depth examination in this project: modular open systems architecture (MOSA), agile acquisition, and rapid prototyping, which are discussed in more detail later in this chapter. While these approaches are not new, AFSPC and SMC have had limited experience in employing these approaches in past acquisition programs,[6] and several new initiatives at AFSPC and SMC are considering these approaches, which we discuss later. The implementation of these approaches in DoD have faced many challenges, but there

---

[5] The Air Force Rapid Capabilities Office was stood up in 2003 with the goal of expediting the fielding of critical capabilities to the warfighter (Randall Walden, "Air Force Rapid Capabilities Office Overview," Air Force Rapid Capabilities Office, June 16, 2016) and has built a reputation of achieving such goals. One of the missions of the ORS Office is to provide space capability to the warfighter in a responsive manner, which includes rapid acquisition (DoD Executive Agent for Space, "Implementation Plan for Operationally Responsive Space," April 2008, Not available to the general public). During preparation of the final report, the ORS Office changed its name to Space Rapid Capabilities Office as directed by the FY 2018 National Defense Authorization Act (NDAA), Pub. L. No. 115-91, § 1601, December 12, 2017, 131 Stat. 1720.

[6] SMC/Program Management and Integration Directorate (PI) staff, discussion on challenges and ways to improve acquisition to support the SEV, El Segundo, Calif., February 8, 2017; past major space programs may have attempted to implement some aspects or elements of MOSA, rapid prototyping, or agile at a smaller scale (e.g., perhaps at the component or subsystem level of development), but based on our discussions with SMC staff, none of the major space acquisition programs were characterized as implementing any of these concepts at a system level, with the exception of JMS Increment 2, which applied an agile acquisition approach (SMC/PI staff, 2017). Early space programs from the 1960s and the 1970s, such as the Defense Space Program, Defense Meteorological Satellite Program, and Defense Satellite Communications System, may have used some variations of elements of MOSA, agile acquisition, and rapid prototyping. However, our research did not examine those programs in detail. Refer to Myron Hura et al., *Space Capabilities Development: Implications of Past and Current Efforts for Future Programs*, Santa Monica, Calif.: RAND Corporation, 2007, Not available to the general public, for details on lessons learned from those programs.

are successful examples. Thus, we judged that they had the *potential* to be successful in the SEV and are worth further exploration.

**Task 2: Characterize challenges to successfully implementing these acquisition approaches.**
MOSA, agile acquisition, and rapid prototyping are versatile acquisition tools, but they differ from a traditional DoD acquisition approach and may face potential barriers or challenges when implementing them in a DoD setting. We identified potential key challenges to successfully implementing these concepts in the SEV based on the following:

1. Examinations of example programs or initiatives, both successful and unsuccessful, that have applied these concepts. We selected the following programs and initiatives because we judged that they shared similar characteristics to some of the programs and initiatives that AFSPC and SMC are considering for the SEV.[7] These programs are discussed in subsequent chapters on MOSA, agile,[8] and rapid prototyping. Given limited project resources, our in-depth examination was limited to the following programs:
    a. A-RCI program (MOSA, successful)
    b. SUMO (MOSA, unsuccessful)
    c. JMS Increment 2 (agile acquisition, unsuccessful)
    d. Air Force Research Laboratory's (AFRL's) Joint Emergent Operational Need (JEON) (agile acquisition, so far successful)
    e. U.S. Navy's Future Naval Capabilities (FNC) program (rapid prototyping, successful)
    f. U.S. Special Operations Command's (USSOCOM's) approach to transition from science and technology (S&T) to acquisition (rapid prototyping, successful)

---

[7] There may be other candidate programs that could provide additional insights, and examining such programs would add value.

[8] Throughout the report, we often substitute the word *agile* for *agile acquisition* when the context allows.

2. Examinations of current practices and processes in space acquisition (including requirements generation) that may constrain application of these concepts.

**Task 3: Identify potential ways to overcome implementation challenges and make recommendations.**
For each of the three promising acquisition approaches that we examined, we identified potential ways to overcome the challenges by distilling lessons from the programs or initiatives examined in Task 2 and best practices from relevant guidebooks. Table 1.1 lists organizations we interviewed for this project. We synthesized findings from our analysis of interviews and literature review to develop an ensemble of detailed implementation recommendations for AFSPC and SMC to consider for the SEV.

**Table 1.1**
**Organizations Interviewed**

| Category | Organizations Interviewed |
|---|---|
| Acquisition professionals | HQ AFSPC/A5/8/9X<br>SMC directorates[a]<br>AFRCO<br>AFRL<br>Air Force Office of Transformational Innovation<br>ORS Office |
| Programs | A-RCI (Navy)<br>Enterprise Space Battle Management Command and Control ([ESBMC2] Air Force)<br>JMS (Air Force) |
| Industry | BAE<br>Boeing<br>Orbital ATK<br>York Space Systems |

[a] Space Superiority Systems Directorate, Remote Sensing Systems Directorate, Global Positioning Systems Directorate, Launch Enterprise Directorate, Military Satellite Communications Systems Directorate, Advanced Systems and Development Directorate, Contracting Directorate, Engineering Directorate, Program Management and Integration Directorate.

**Task 4: Apply these recommendations to the Enterprise Space Battle Management Command and Control program.**
The ESBMC2 program is intended to be a critical enabler to realizing the SEV, as we will discuss in Chapter 5, and it is considering the use of MOSA, agile, and rapid prototyping for its acquisition.[9] Per our sponsor's guidance, we applied our findings to the ongoing ESBMC2 program as a concrete application of this research.

### Research Limitations

Our analysis is an initial look at how these approaches can be applied. There are many barriers and challenges to the full and successful application of these approaches, and they may vary depending on the specific program. Further, many other variables that are independent of these approaches may affect the success of a program. Thus, it was beyond the scope of our research to assess the likelihood of successful implementation of these approaches.[10] Instead, our research focused on identification of potential challenges so that they can be anticipated and managed to improve the likelihood of success should these acquisition approaches have a role in the SEV. Given limited project resources and evolving space architectures to support the SEV, our research did not fully examine the applicability of these approaches to a range of space acquisition programs that are in the Air Force's portfolio; nor did we conduct a comprehensive risk analysis to assess potential risks of employing or adopting these approaches. The research findings are based on data collected through August 2017.

---

[9] SMC, Space Defense Task Force, "Space Defense Task Force All Call," briefing, May 2017; SMC, "Maturation of Critical Technologies for Space Battle Management Command and Control (BMC2)," *Broad Agency Announcement (BAA)*, FA8819-17-BMC2-BAA, May 22, 2017.

[10] Although MOSA, agile, and rapid prototyping have the potential to improve the acquisition and fielding timelines, we recognize that there are a large number of variables that affect the program outcome, and we do not evaluate such variables (e.g., funding stability, adequate cost estimates, and adequate systems engineering, among others). We did not evaluate such programs that have attempted to apply MOSA, agile, or rapid prototyping to understand the causes of the program failure because there may be other factors that contributed to program difficulties.

## Overview of Acquisition Approaches

### Modular Open Systems Architecture

A MOSA system is characterized by a modular design with loosely coupled, plug-and-play system modules that are connected via widely supported, consensus-based interface standards.[11] While there are many potential benefits of a MOSA approach, the two primary benefits that are most applicable to our research are *rapid upgradeability* and *enterprise synchronization*.[12] MOSA enables capability development and upgrade at a lower functional level without system redesigns or system replacement. As a result, MOSA offers opportunities for more frequent technology insertions to enable responsiveness to rapidly changing capability needs as the threat evolves. Thus, in practice, MOSA is implemented to allow incremental development and fielding of capabilities. Further, MOSA enables technology and module reuse across an enterprise, which not only reduces the need for unique development efforts, eliminating significant development costs, but can enable improvements in interoperability across the enterprise.[13] While these

---

[11] The International Telecommunications Union defines "open" standards as "standards made available to the general public and are developed (or approved) and maintained via a collaborative and consensus driven process" (International Telecommunications Union, *Definition of "Open Standards,"* November 11, 2015). We define "proprietary" or "closed" standards, then, to be standards that are managed by a single entity through internally governed processes that do not obtain any level of industry consensus. In some cases, proprietary standards may be made available for use by third parties but with limited design flexibility and potentially at a high cost. See U.S. Department of Defense, *Program Manager's Guide to Open Systems*, Version 2.0, Washington, D.C., September 2004; and Office of the Deputy Assistant Secretary of Defense-Systems Engineering, "Initiative: Modular Open Systems Approach," webpage, undated.

[12] Other benefits include enabling the broad use of competition in system procurements, facilitating technology refresh, increasing flexibility and innovation, and increasing cost savings from technology reuse and shared investment. Appendix A in the online appendixes further discusses key attributes and benefits of MOSA in detail.

[13] Interoperability is defined by DoD Directive (DoDD) 5000.01 as "the ability of systems, units, or forces to provide data, information, materiel, and services to and accept the same from other systems, units or forces, and to use the data, information, materiel, and services so exchanged to enable them to operate effectively together." See U.S. Department of Defense, *Directive 5000.01: The Defense Acquisition System*, Washington, D.C., November 20, 2007.

benefits are desirable, MOSA is not applicable to all situations. Due to the depth of DoD guidance encouraging the use of MOSA in a variety of applications, the goal of "openness" or "open architecture" is often pursued as an end, when it should be considered a means to an end: achieving the benefits discussed earlier. In fact, it does not make sense to use MOSA systems in all applications, and there are situations where a closed system may be more appropriate than an open one.[14]

Unfortunately, MOSA is often forced into situations where it is not suited based on a desire to achieve its benefits without understanding what is required to make it successful. Agencies wishing to implement MOSA should evaluate their circumstances to determine if MOSA makes sense for their application and define programmatic end goals associated with the implementation of MOSA early in a program's life to prevent commitment to a design-and-acquisition approach that is not appropriate.[15]

### Agile Acquisition

Agile acquisition arose (most recently) from the application of Lean Six Sigma methods to software development.[16] In recent years, it is increasingly being applied to hardware development as well. While

---

[14] One example might be instances when there is not a strong reason to make subcomponents separately procurable (e.g., only a single vendor may be qualified to provide specific products). Another example might be a situation where no suitable standard exists and existing standards are not widely adopted enough and too immature to consider for the application (Donald Firesmith, "Open System Architectures: When and Where to Be Closed," *SEI Insights Blog*, Carnegie Mellon University, Software Engineering Institute, October 2015).

[15] Carol Sledge, "A Discussion on Open Systems Architecture," *SEI Insights Blog*, Carnegie Mellon University, Software Engineering Institute, November 23, 2015.

[16] For an overview of the management paradigm shift that Lean Six Sigma emerged from, see Arnold Levine and Jeffrey Luck, *The New Management Paradigm: A Review of Principles and Practices*, Santa Monica, Calif.: RAND Corporation, MR-458-AF, 1994. It also shares some important attributes with iterative and incremental development. For a brief history, see Craig Larman and Victor R. Basili, "Iterative and Incremental Development: A Brief History," *Computer*, Vol. 36, No. 6, IEEE Computer Society, June 2003, pp. 47–56. Agile shares attributes with earlier approaches to evolutionary and spiral development but has moved well beyond them in the processes, methods, tools, and skills that it applies.

a wide range of agile practices exist, they tend to share a number of characteristics:[17]

- incremental approach: incremental development with rapid testing during/after each increment
- flexible approach: adjusting requirements repeatedly as increments accumulate
- user input: giving users usable product for rapid, hands-on assessment
- tailoring: delivering only product that users need, including documentation
- delegation: empowering smaller, higher-skilled teams to do these things with fewer formal processes.

Although agile acquisition is being extended to the development of hardware, with or without closely associated software, experience remains deepest in its application to software. As a result, the net benefits of its application tend to be higher in software than in hardware development. Experienced users who now prefer the use of agile acquisition for almost any software development effort are typically more cautious when they consider using it in hardware development. A motivating principle behind agile acquisition is that it is simply impossible to anticipate what specific capabilities users will need in the future. This principle becomes more important when the environment in which a system will operate becomes more dynamic. As time passes during development, surprises will arise. These can include new threats in space or unexpected failures of the technologies already present in space. They can include the unexpected emergence of new technologies or sources of technology. Agile acquisition anticipates that such surprises will occur. Developers cannot know in advance which will occur; however, the agile process enables developers to react rapidly when they inevitably do.[18] To the extent that SEV seeks to improve

[17] Appendix C discusses these characteristics in more detail.

[18] A more traditional waterfall approach can of course also react to such surprises. But because it involves much less delegated authority, involves more staff and decisionmakers, and may not prioritize quick response relative to other development goals, traditional bureaucratic processes often slow effective responses.

the robustness of our space capabilities when space is contested, agile acquisition may be a promising way to accomplish such goals.

### Rapid Prototyping

Prototyping is typically characterized as a set of design-and-development activities intended to reduce technical uncertainty and to generate information to improve the quality of subsequent acquisition decisionmaking. It is traditionally accomplished in the early stages of an acquisition program of record (POR) to reduce technical risk prior to proceeding into initial production and the operations and support (O&S) phase of an acquisition program. In our research, we focus on a narrow application of prototyping based on our sponsor guidance and interest. That is, we define *rapid prototyping* as experimentation conducted outside the POR to field a capability faster than a traditional acquisition POR.[19] The primary goals of AFSPC's rapid prototyping efforts differ from those of traditional prototyping. They are

- rapidly operationalizing and deploying an "as built" prototype[20]
- conducting rapid procurement and fielding of additional prototype systems/capabilities
- rapidly transitioning the prototype(s) to an acquisition POR for enduring operational support and sustainment of the capability.[21]

As AFSPC develops new space architectures and operational capability requirements to implement the SEV, we foresee three circumstances in which AFSPC may want to conduct a rapid prototyping activity:

- to respond rapidly to technological surprise by an adversary that threatens space systems and capabilities

[19] AFSPC staff, teleconference discussion on prototyping, May 17, 2017. The primary purpose of AFSPC's rapid prototyping activity is not for addressing urgent operational needs, though it can.

[20] In this context, "as built" means minimal modifications would be made to the prototype after the experimentation and prior to deployment.

[21] The intent is to select prototypes that require minimal development after the prototype transitions.

- to implement planned enhancements to space architectures and systems
- to demonstrate and evaluate promising, innovative technologies in an operational setting.

AFSPC is currently developing and implementing a process to nominate candidates for rapid prototyping.[22] In support of this effort, AFSPC asked PAF to identify ways to *rapidly transition* from prototype demonstration and evaluation to production and operational deployment.

**Opportunities to Apply Modular Open Systems Architecture, Agile Acquisition, and Rapid Prototyping in the Space Enterprise Vision**

While MOSA, agile, and rapid prototyping may be implemented independently, there are instances in which they may be used together synergistically. For example, within a MOSA framework, when properly designed, development and upgrades of each module can proceed independently. Each module can be developed and upgraded using an agile or traditional waterfall approach or rapid prototyping. In addition, rapid prototyping could occur in an agile acquisition environment. Agile acquisition lends itself to rapid prototyping because, by design, each increment in agile acquisition yields a usable product, which could be a usable prototype. Meanwhile, the prototype can generate data based on operational experience that users can bring back to developers to help decide which capabilities to pursue in future increments.

In principle, these approaches could be applied to the space or the ground segments of the space enterprise for hardware or software systems. However, near-term opportunities appear to be in the ground segment. AFSPC and SMC's Enterprise Ground Services (EGS) initiative is developing a MOSA architecture and associated standards and interfaces to enable the transition of legacy ground systems and build future ground systems to a common ground architecture, which we will

---

[22] AFSPC is determining what to prototype based on operational gaps or shortfalls. It is currently working to codify a prototyping process within an AFSPC instruction (AFSPC staff, 2017).

briefly discuss in Chapter 2.[23] The ESBMC2 program is considering MOSA, agile, and rapid prototyping in its acquisition, which we will discuss in Chapter 5. At a smaller scale, the Global Positioning System (GPS) directorate at SMC is considering an agile approach for software upgrades and modernization in the sustainment portion of the GPS ground system.[24] Similarly, there may be opportunities to apply agile methods and prototyping for software applications on a smaller scale for other sustainment and modernization programs at SMC.[25]

Applying MOSA, agile, and rapid prototyping for the space segment does face some limitations, at least for hardware. One of the primary objectives of these concepts is to enable more frequent capability upgrades (in many cases, in an incremental manner). However, limited physical access to deployed space assets greatly hampers in situ hardware upgrades to space segment hardware.[26] Capability upgrades could be implemented on subsequently produced and deployed units, but current space systems are designed with a long lifetime (typically over ten years).[27]

That said, the SEV envisions future space architectures with more proliferated satellite constellations and more frequent constellation

---

[23] SMC/AD staff, discussion on challenges and ways to improve acquisition to support the SEV, El Segundo, Calif., January 20, 2017.

[24] SMC/GP staff, discussion on challenges and ways to improve acquisition to support the SEV, El Segundo, Calif., February 8, 2017.

[25] One of the SMC directorates is considering prototypes and agile methods for mission data exploitation software (SMC/RS staff, discussion on challenges and ways to improve acquisition to support the SEV, El Segundo, Calif., February 7, 2017). While the prototyping work is occurring within the POR, this example illustrates that rapid prototyping that AFSPC envisions could have a broader application to include software sustainment and modernization.

[26] Programmable hardware, such as software-defined radio, could be upgraded on orbit. On-orbit satellite servicing may enable in situ hardware upgrades to satellites in the far term as relevant technologies, operational concepts, and policies mature.

[27] Other than the GPS constellation, which nominally consists of 32 satellites, other DoD constellations are much smaller. The Air Force has procured six Space Based Infrared System (SBIRS) GEO and six Advanced Extremely High Frequency (AEHF) satellites since 1996 and 2001, respectively (Kim et al., *Acquisition of Space Systems: Past Problems and Future Challenges, Volume 7*, Santa Monica, Calif.: RAND Corporation, MG-1171/7-OSD, 2015).

upgrades (which may imply shorter satellite life) to increase resiliency. Additionally, AFSPC's initial rapid prototyping activities are likely to include prototypes for space hardware, possibly leveraging the long-duration propulsive-evolved Evolved Expendable Launch Vehicle secondary payload adapter space platform or hosted payloads for space situational awareness or space protection applications.[28] As the future space architectures become fleshed out, AFSPC and SMC could look for additional opportunities where these approaches could provide value.

## Organization of This Report

This report is organized as follows. The next three chapters discuss MOSA, agile acquisition, and rapid prototyping in turn. Each chapter provides an introduction to these promising acquisition approaches, challenges to carrying them out in a DoD setting, and recommendations for implementation. Chapter 5 applies the insights from the earlier three chapters to the ESBMC2 program. Chapter 6 summarizes the findings and recommendations in this report.

In addition, six online appendixes provide background and reference information on MOSA, agile acquisition, and rapid prototyping to supplement our analysis. Appendixes A and C, in particular, contain background information on MOSA and agile, respectively, for those readers who may not be familiar with these concepts. The appendixes are available for download at www.rand.org/t/RR2626.

---

[28] AFSPC staff, 2017.

CHAPTER TWO

# Modular Open Systems Architecture

Implementing MOSAs in the SEV offers the potential to upgrade capabilities more responsively and enable enterprise synchronization. While the concept of MOSA is relatively straightforward, its implementation within the DoD acquisition management framework requires a much deeper understanding and appreciation of the complexities associated with it. There are several key challenges to successfully implementing a MOSA in DoD, and this chapter discusses those challenges and potential ways of overcoming them to improve the likelihood of success should MOSAs have a role in the SEV. This chapter first begins with a discussion of example MOSA programs to provide additional context for our analysis.

## Applications and Limitations of Modular Open Systems Architecture in the Space Enterprise Vision

MOSAs have been applied to both hardware and software in DoD. One notable successful example is the Navy's A-RCI program, which began in the 1990s, to incrementally upgrade and field the sonar processing system (hardware and software) across all submarine classes using a MOSA approach.[1] The program's goal was to create an evolving sonar processing system with capability improvements delivered

[1] The A-RCI program is discussed in more detail in Appendix B in the online appendixes.

as fast as commercial technology became available to counter increasing threats. Another goal was to bring commonality to the fleet to potentially reduce maintenance and training requirements. A-RCI is currently installed on all in-service U.S. Navy submarines and therefore requires an enterprise entity to develop its operational requirements with all ship classes in mind. A-RCI demonstrates that situations where system capability can be decomposed into smaller parts to take advantage of incremental system fielding and upgrades are a place where a MOSA makes sense. A-RCI also demonstrates that MOSA makes sense to use in situations where component commonality across a large enterprise can be leveraged. Justification for shared infrastructure or common standards may be limited in a situation where only a low volume of components is being procured. However, if commonality can be enacted for applications with similar design requirements or constraints, economies of scale can be achieved.[2] The ground enterprise in the SEV presents this opportunity as we describe below.

The Air Force's current ground architecture consists of multiple, disparate ground systems for satellite command and control (C2), mission planning, mission data processing, and other support activities. Each ground system is customized and dedicated to a particular satellite constellation or satellite program.[3] When a new satellite or constellation is acquired, a new, dedicated ground system is acquired along with it. Although these ground systems possess some common functionalities and infrastructure, they are not interoperable. Having some degree of interoperability could enable one ground system to provide back-up capabilities if another is down, thereby increasing the resilience of the ground enterprise. Because of potential inefficiencies and the desire for increased interoperability in the ground architecture,

---

[2] Commonality and modularity need to be designed and selected judiciously, balancing potential benefits and risks. While these design attributes can reduce costs, those cost savings could be offset if, by introducing commonality and modularity, other development risks and costs are incurred elsewhere.

[3] For example, the ground systems for SBIRS, GPS, Wideband Global SATCOM, and AEHF are unique to each constellation and are separately acquired for each satellite program.

AFSPC and SMC initiated the EGS to enable the transition toward a common ground architecture.[4]

The goals of EGS are to enable the sharing of common infrastructure, interoperability of ground systems across the enterprise, scalability of the ground system as additional satellites are added rather than building a new ground system from scratch, and rapid upgradeability as new capabilities become available.

The concept of a common ground system for disparate platforms is not new. The Air Force's MMSOC is currently using a common ground system based on a MOSA for control and operations of disparate research and development (R&D) satellites that support different missions.[5] EGS plans on building upon MMSOC ground system architecture. The Navy and the National Aeronautics and Space Administration (NASA) have also developed common ground architectures based on MOSAs to enable commonality and scalability in ground systems to support multiple disparate satellites.[6]

Similarly, on the airborne side, AFRCO's Common Mission Control Center (CMCC) program uses a MOSA-based common ground architecture for control and operations of disparate Unmanned

---

[4] Initial efforts on EGS appear to focus on developing a MOSA for software systems to enable life-cycle software upgradeability (SMC/AD, "Enterprise Ground Services," briefing, undated). A push for commonality and expanded use of COTS/government off-the-shelf (GOTS) components when possible in ground station hardware could lead to savings in nonrecurring engineering and hardware procurement costs (due to economies of scale) for the ground enterprise.

[5] MMSOC also supports Space Based Space Surveillance system and has supported operational satellites that have transitioned from R&D. MMSOC began operations in 2010 (Tiffany Morgan, "Multi-Mission Satellite Operations Center Ground System Architecture (MMSOC GSA)," SMC/SDTC, February 29, 2012).

[6] The Navy uses a government-owned Neptune Common Ground Architecture software system for operations of more than 80 different satellites for the Navy and other government agencies (Daniel Parry, "NRL at Blossom Point Celebrates 60 Years of Service," *The U.S. Naval Research Laboratory: News Releases*, June 14, 2016). NASA Goddard Space Flight Center developed Goddard Mission Service Evolution Center architecture, which relies upon an application programming interface definition, standard message definitions, and communications middleware to implement the architecture. The architecture and its components are implemented across the NASA centers and other government organizations (NASA, *Goddard Tech Transfer News*, Vol. 9, No. 2, 2011, pp. 4–5).

Aircraft System (UAS) platforms. CMCC is based on the Open Mission Systems (OMS) architecture and Unmanned Aerospace Systems Command and Control Initiative (UCI) standards for the C2 and mission management system to enable interoperability across disparate UAS C2 systems and to enable software reuse across different C2 systems.[7] As we will discuss in Chapter 5, AFRCO is working on expanding OMS/UCI for application to space BMC2.

Use of MOSA in SEV applications beyond the ground segment brings challenges. MOSAs have been attempted for satellites but have had limited success. The ORS Office attempted a MOSA approach to enable rapid integration of bus components and integration of payloads to buses to support the ORS mission; it pursued establishing bus standards and bus/payload interface standards.[8] However, this approach did not fully materialize, partly because standard ORS buses would be procured in low quantity.[9] Subsequently, the Office of the Director of National Intelligence led a broader effort called SUMO to establish common bus standards for the U.S. space industry to serve commercial, civil, and national security space customers.[10] In the end, SUMO also failed to achieve a set of standards that the space industry can adopt.

There were many factors that made achieving common bus standards difficult, and we highlight a couple of key issues here.[11] For one,

---

[7] AFRCO, "Common Mission Control Center (CMCC) Program Overview," May 16, 2017, Not available to the general public.

[8] ORS Office staff, teleconference discussion on ORS lessons on responsive acquisition, March 9, 2017; Brian Davis, "Interface Standards as an Enabler for Operational Responsive Space," presented at the Ground System Architectures Workshop, Redondo Beach, Calif., March 31–April 3, 2008.

[9] For instance, since the ORS Office stood up in 2007, it has procured four ORS satellites through 2017, and not all of them used a standard ORS bus.

[10] Bernie F. Collins, "Development of a Space Universal Modular Architecture (SUMO)," 2013 IEEE Aerospace Conference, Blue Sky, Mont., March 2–9, 2013a.

[11] Boeing staff, discussion on MOSA, agile, and rapid prototyping, El Segundo, Calif., June 29, 2017; Orbital ATK staff, discussion on MOSA, agile acquisition, and prototyping, Arlington, Va., June 27, 2017; Bernie F. Collins, "Space Universal Modular Architectures (SUMO): Commercial Space Industry's Dilemma," June 2013b.

the commercial space industry was not convinced that transitioning to common standards would provide benefits without an active government role. For instance, the primes already have their own line of common buses, and thus they were concerned that nonengineering costs associated with adopting a different set of standards would lead to reduced profits (especially since they were not confident that there would be a government program that could offset the nonrecurring engineering costs).[12] Second, there was no government funding for industry participants to form a consortium for selecting or developing industry consensus-based, common standards. Standards development or selection would be an intensive effort that could take several years.[13] One of the key insights from these initiatives is that the primary barrier of MOSA implementation for satellites was a business case issue more so than a technical issue.

## Modular Open Systems Architecture Implementation Challenges

While such programs as A-RCI and CMCC indicate MOSA is possible in a DoD acquisition environment, MOSA implementation is not without challenges. We examined several programs or initiatives that attempted to implement a MOSA and interviewed acquisition officials and industry subject-matter experts to gain insights into key challenges they faced. These challenges included

- aligning diverse stakeholders when MOSA is applied to a large enterprise

---

[12] A related issue is that the volume of satellites produced in the space industry (commercial and government combined) may not be sufficiently high to warrant standardization. CubeSats, however, are an example of successful standardization in the space industry. They are produced and launched in much higher volume than the traditional commercial and government satellites.

[13] SUMO's notional plan for developing standards was estimated to take at least six years (Collins, 2013b).

- managing acquisition using incremental development and fielding
- determining key interfaces and the appropriate standards for those interfaces
- tailoring industry engagement to develop consensus on standards and to acquire data rights, if needed
- selecting the appropriate systems integration model.

We discuss these challenges in turn in this section.

### Aligning Diverse Stakeholders

Common interface standards for a broad enterprise enable interoperability and technology reuse across the enterprise, but their acceptance and adoption require the agreement of diverse stakeholders on common goals and requirements. These stakeholders include individual program managers responsible for programs within that enterprise. Because an individual program's success focuses on achieving specific performance, cost, and schedule goals for a given system, achieving technical success at the portfolio or enterprise level may involve making nonoptimal program-level decisions for the greater good of the enterprise or portfolio.[14] Obtaining consensus at the enterprise level is challenging because individuals at lower levels of governance are often evaluated based on the success of achieving program-level objectives and not the overall performance of the portfolio/enterprise. As a result, these individuals do not have inherent incentives to put the needs of the enterprise first. Based on discussions with various Air Force program managers, we observed that space system development and procurement are often managed at the program level. This is not to say that no entity considers higher-level space operational needs, but rather optimization of individual space systems is the primary focus of space program managers. As an example, each satellite program at

---

[14] At the portfolio level, trade-offs are balanced across multiple programs to achieve a common capability goal. The enterprise level involves a combination of interoperable portfolios, potentially at the department or service level (MITRE, *Systems Engineering Guide*, Bedford, Mass., 2014).

SMC procures a customized bus for the individual program.[15] SMC has explored the potential applicability of a common bus that could be designed and developed for multiple programs to reap potential cost and schedule savings (through block buys and reduced nonrecurring engineering).[16] However, this concept did not gain traction because a common bus implied a design that is not optimal for any one program; a common bus could potentially yield additional requirements that are not relevant to some programs, or if compromises had to be made, the bus performance may be less than optimal for other programs.[17] In the current acquisition management construct at SMC, a program manager would not be incentivized to compromise individual program goals for the benefit of the enterprise.

The SEV would be a major step toward taking an enterprise view for space acquisition; however, establishing detailed directives and enforcement for issues like commonality and accomplishment of an operational need by multisystem interoperability are necessary to keep diverse stakeholders in alignment. In addition to the establishment of these directives, DoD may need to establish a single entity responsible for these issues across key portfolios or even across the enterprise.

### Managing Acquisition with Incremental Development and Fielding

Traditional DoD acquisition processes and time lines involve the achievement of a linear set of milestones by a program. These milestones include the establishment of capability requirements documents based on operational needs. MOSAs are designed to take advantage of flexible, modular architectures to quickly and incrementally develop

---

[15] Although many of the satellites that SMC procures are based on a "common bus" that each manufacturer uses (e.g., Lockheed Martin's A2100 bus), they are modified to meet an individual program's needs.

[16] SMC/AD staff, 2017.

[17] Each satellite constellation operates in a different environment and serves a different mission, which imposes different requirements on the bus in terms of parts qualification and size, weight, and power, among other requirements. For example, strategic satellite systems, such as SBIRS, GEO, and AEHF, have more stringent hardness requirements than nonstrategic systems because they need to be nuclear survivable.

capability improvements, something that is challenging to do when governed by a rigid, linear acquisition process.

Fortunately, recent emphasis on tailoring and streamlining enables and even encourages programs to be set up incrementally to take advantage of the benefits of an approach like a MOSA.[18] Successful historic examples of incrementally implemented programs are few, for such reasons as using thresholds or goals to define requirements that can change over time; including design margins to ensure room for capability growth in future system increments; and establishing agreeable test and evaluation plans to demonstrate incremental or partial system performance. The A-RCI program, discussed in more detail in Appendix B of the online appendixes for this report, is an example of a successful incremental program. A-RCI has established a calendar-driven program schedule that incrementally satisfies acquisition process guidelines (e.g., scheduled, periodic Joint Capabilities Integration and Development System [JCIDS] reviews and periodic integrated developmental and operational tests) while still making speedy technology insertions into operational platforms to meet evolving operational needs. Taking advantage of the dispersed DoD expertise on how to design and implement incremental programs successfully is a challenge to the successful implementation of MOSAs.

### Determining Key Interfaces and Appropriate Interface Standards

While there are benefits to interface standards, it is important to note that not every interface involved in connecting system elements must be standardized. MOSA system owners must determine which *key* interfaces should be standardized and governed by an interface standard and which do not need to be. Interfaces across which interoperability is required and modular development will be practiced are the key interfaces that must be managed via standards. Once the key inter-

---

[18] Including multiple approval and documentation models for incrementally approved programs and a specialized "IT [information technology] Box" model for IT systems based on COTS/GOTS hardware and incremental software deployments. See DoD, *Operation of the Defense Acquisition System*, Washington, D.C.: Department of Defense Instruction 5000.02, Change 2, February 2, 2017; and MITRE, *Acquisition in the Digital Age: Analyze Requirements*, undated.

faces are selected, MOSA system owners must select an appropriate set of standards to govern the key interfaces.

Selection of appropriate standards must be based on technical need, but also maturity and level of market adoption. As is discussed in further detail throughout this chapter, the use of mature, widely adopted standards where possible benefits a system's upgradeability, interoperability, and life-cycle cost.[19] If only narrowly adopted or immature standards exist for an interface, forcing the use of such a standard could result in challenges for a system's life-cycle support efforts. This is because only a select few suppliers would be available to participate in component development competitions. Standards management and revision is a key aspect of using interface standards for MOSAs (see Appendix A in the online appendixes on standards management). That said, there are mechanisms for DoD to sponsor the development of a standard for a given enterprise in cases where no suitable standard exists but one is desired.[20] Such an effort should not be taken lightly, because ensuring a standard's success involves significant long-term management and support. Developing a commercial standard from the ground up is a complex process with many stakeholders, and finalizing the standard can take a long time.[21] Tying a system's development to the completion of a new standard that is expected to take a long time could result in delays to that system.

### Tailoring Industry Engagement on Interface Standards and Data Rights

If multiple mature, widely adopted standards exist, selecting a single standard can be a challenging proposition. In such an environment, selecting any available standard would potentially leave certain suppliers out of development discussions unless they adopt the selected standard. Based on discussions with multiple space industry entities, it appears that this situation is present in the space environment. Current

---

[19] U.S. Department of Defense, 2004.

[20] One such process, U.S. Department of Defense, Defense Standardization Program (DSP) Procedures, DoD 4120.24-M, December 2017.

[21] Sledge, 2015.

divisions between commercial and government space markets, as well as limited procurement volumes for given space customers, have hampered voluntary standardization within the space community. In the absence of a driving force with large market share, achieving standards commonality in space could continue to be challenging. When industry does not agree on common standards, the number of suppliers that can participate in a given subcomponent development is smaller, and achieving the full benefits of the MOSA becomes more challenging.

One of many benefits of MOSA is that it enables system designers to consider a broad supplier base for each module and, in some cases, to recompete individual subcomponent contracts.[22] To reopen production competition for a component already in production, a design agent must own sufficient data rights to allow another supplier to produce the given component. This is known as form, fit, and function data. Additionally, for a third-party integrator to integrate all of a MOSA's subcomponents into a single operational system requires a detailed understanding of those subcomponents. If the system integrator is not the same as the lead system agent, the system agent must own sufficient data rights to enable the integrator to perform their function successfully. Data rights issues are discussed in more detail in Appendix A.

### Selecting the Appropriate Systems Integration Model

Finally, in a DoD setting, two basic integration models exist for MOSAs, depending on which organization will be responsible for system integration: either the government or a contractor. This decision is somewhat specific to MOSA because, as identified above, traditional developments often involve a monolithic model with proprietary interfaces and subsystems wherein a single entity (almost always a prime contractor) designs the system, develops or subcontracts the subcomponents, and integrates them into a full system before delivery. Sharing this proprietary data from one contractor to another may

---

[22] In these cases, the DoD should assess whether the costs associated with obtaining technical data rights is outweighed by the potential benefits of pursuing alternative sources.

be unpalatable if the two organizations compete on other contracts. Since the government is not competition for contractor design agents, using a government integrator could be one way to avoid this conflict. Unfortunately, based on our conversations with multiple government and industry entities, government organizations often lack the deep technical knowledge needed to carry out complex system integration.

Under a MOSA, a system integrator ideally would only need to know the system architecture and standards. If the design was performed by a different design agent, the overall architecture and design would need to be provided but not the design specifics within the component modules. However, some level of more detailed design engagement could be required between component suppliers and system integrators.[23] This relationship should be considered early in a program's life. In the case of A-RCI, while individual software algorithms are procured competitively, the Navy has a long-term relationship with a single integration vendor.

## Overcoming Challenges to Implementing Modular Open Systems Architecture

We distilled lessons and best practices from the A-RCI program case study, discussions with subject-matter experts, and guidebooks on MOSA applications in DoD to identify potential ways to overcome the identified challenges and improve the likelihood of success where MOSA may have a role. We summarize these findings in terms of three overarching recommendations:

---

[23] Such integration challenges were raised in our interviews (Boeing staff, 2017; A-RCI Program Office staff, discussion on A-RCI lessons on MOSA implementation, Washington, D.C., January 31, 2017; Air Force OTI staff, discussion with authors on OTA, Open Architecture, Arlington, Va., May 31, 2017b) and in the JCREWS program (Frank Camm, *Analysis of Technology Innovator Industrial Base for Joint Counter Radio-Controlled Improvised Explosive Device Electronic Warfare 3.3*, Santa Monica, Calif.: RAND Corporation, 2011, Not available to the general public). Appendix A provides additional information on data rights issues for integration in MOSA.

- establish an enterprise governance
- invest in up-front systems engineering and infrastructure
- develop up-front life-cycle business and program plans.

To be successful in implementing MOSA, the first overarching recommendation and its supporting elements must be enacted before proceeding to the other two. The second and third overarching recommendations then must be considered in conjunction with each other. Table 2.1 summarizes these recommended activities, indicating whether they apply to AFSPC or SMC.

### Overarching Recommendation 1: Establish an Enterprise Governance

An environment with an established enterprise, shared goals, and enterprise-level governance helps decisionmakers make trade-offs

**Table 2.1**
**MOSA Recommendations for AFSPC and SMC to Enable SEV Goals**

| Recommendation | AFSPC | SMC |
|---|---|---|
| *Overarching Recommendation 1: Establish an Enterprise Governance* | | |
| Define the enterprise scope and relationship to organizational goals and incentives | X | X |
| Provide governance members the data needed to inform decisionmaking | X | X |
| Obtain executive support and formalize metrics-based evaluation processes | X | X |
| *Overarching Recommendation 2: Invest in Up-Front Systems Engineering and Infrastructure* | | |
| Develop a functionally partitioned design | | X |
| Select appropriate interface standards | | X |
| *Overarching Recommendation 3: Develop Up-Front Life-Cycle Business and Program Plans* | | |
| Develop business engagement plans to enable competition | | X |
| Develop integrated life-cycle program plans | | X |

among the needs of diverse stakeholders. Enterprise-level governance is a set of responsibilities and practices exercised by a board of executive management to provide strategic direction, ensure objectives are achieved, ascertain that risks are managed appropriately, and verify that the enterprise's resources are sufficient and used responsibly.[24] To succeed, the enterprise governance needs to be hierarchical to be able to delegate certain responsibilities to the lower levels of governance discussed earlier in this report. It also needs to understand the technical trade-offs within the enterprise to make informed decisions.

In the case of A-RCI, enterprise governance was provided by the cooperative efforts of the Submarine Tactical Requirements Group (STRG), which is a group of senior submarine community leaders who represent the user community, and the Program Office. Using a combination of operational, technical, and business expertise, these organizations established an initial time line and fielding schedule for their system. For future iterations, their collaborative efforts defined the technical content of a fielding increment and the business approach to procure that military capability.[25]

Based on our analysis, elements we assessed as key to the SEV context are

- goals and outcomes tied to the organization's goals and incentives
- sufficient data to make informed decisions
- sponsorship by senior leadership
- a formal structure with defined and repeatable processes, as well as performance metrics to measure success.

#### *Define the Enterprise Scope and Relationship to Organizational Goals and Incentives*

The first supporting step to defining an enterprise governance program is establishing what will be governed by the program and how the governance structure will meet the organization's overarching goals. Tying the governance goals to organizational goals and appropriate

---

[24] MITRE, 2014.

[25] Discussed in more detail in Appendix B.

incentives is key to mitigating the risk that program-level governance becomes solely focused on achieving its own success at the expense of the enterprise.[26] This narrower focus undermines the successful implementation of MOSA.[27] In the SEV context, this step includes the detailed delineation of what systems must be compatible, what systems must be interoperable, or what systems should share technology commonality—systems in the space segment, systems in the ground segment, or systems from particular mission areas, and so on. Once this is established, specific technical solutions and interface standards may be considered to meet those needs. Importantly, this effort must include a connection to the Air Force's space strategic goals. All levels of this enterprise governance structure must share common goals in line with the SEV goals.

#### *Provide Governance Members the Data Needed to Inform Decisionmaking*

Members of the enterprise governance must be knowledgeable enough to understand operational, technical, and business trade-offs associated with decisionmaking within the enterprise. However, those individuals cannot always be an expert in each of those topic areas. They must have a trusting relationship with lower-level elements of staff and governance in the enterprise to make informed decisions. Though the enterprise governance should have a trusting relationship with individual stakeholders, by design, its members are best positioned to consider the inputs of all stakeholders and make a decision in the collective best interest in the event of conflict.[28] In the case of A-RCI, interchanges between the STRG and technical arm of the program office drove focus areas for technology development along an established, calendar-driven time line.[29]

---

[26] Further research on how to structure the incentives to align with organizational goals would be value added.

[27] MITRE, 2014.

[28] MITRE, 2014.

[29] See Appendix B.

A potential MOSA for a future Air Force space program would include the development of capability requirements for an increment of capability, weighing in on the appropriateness of a commercial standard for a particular application, determining if a new standards development effort is feasible or appropriate, providing recommendations or guidance for industry involvement in an engagement, or establishing a program time line—all of which are detailed issues that must be resolved early in a program's life.

Achieving this agenda in the SEV context would involve membership from AFSPC and SMC. For example, in the case of requirements development, a single entity with expertise in technical issues could be given responsibility for developing cross-enterprise requirements. This body should regularly interact with operational communities to determine requirements and acquisition communities to determine realistic procurement budgets and time lines. Technically speaking, this involves defining high-level requirements that are refined in greater detail for potential capability increments. These early, top-level[30] system designs must include such flexibility tools as margins to provide room to grow over the life of design (a major motivation for pursuing MOSA) and mitigate program risks. From a DoD programmatic process perspective, program managers must obtain Joint Requirements Oversight Council approval for their incremental requirements and obtain Milestone Decision Authority approval for their proposed tailored incremental strategy and associated oversight adjustments according to the DoD Instruction (DoDI) 5000.02.

### *Obtain Executive Support and Formalize Metrics-Based Evaluation Processes*

Obtaining the support of an organization's executive leadership and implementing formal processes with measurable metrics are related elements of establishing enterprise governance. These activities legitimize the organization to mitigate the likelihood of enterprise members attempting to bypass the governance structure to achieve their own

---

[30] A top-level design is a projection of how a system might look to meet planned operational requirements. See MITRE, 2014.

goals. While bypassing the enterprise governance may benefit their individual goals, such an action could have significant long-term costs and undermines the purpose of the governance structure. Formalized processes with full support of senior executive leadership are less likely to be bypassed.[31] In the context of the SEV, formalized oversight and metrics processes should focus on the larger enterprise goals as discussed above but could also include operational considerations, such as system requirements development, industrial engagement, and system life-cycle management as they pertain to enterprise objectives. Tracking metrics and tying them to organizational incentives could help drive enterprise-focused behavior.

### Overarching Recommendation 2: Invest in Up-Front Systems Engineering and Infrastructure

Once an enterprise scope is defined and governance structure established, AFSPC and SMC can begin to consider the technical aspects of implementing MOSAs.

#### *Develop a Functionally Partitioned Design*

Using the MOSA principles, SMC—with consideration for enterprise interoperability needs directed from AFSPC—can look for places to implement a modular design within the space enterprise. As discussed earlier, this appears more straightforward for the ground segment. C2 capabilities implemented on relatively standard computer hardware can themselves be implemented as a modular hardware system following the A-RCI model. Ground-segment software can similarly be modular, provided there is a standardized software infrastructure or use of middleware.

Applying MOSA to the space segment may be more challenging. A software approach involving a standardized software interface similar to the one discussed above could be employed to enable faster software upgrade time lines. This may require spare onboard processing capacity in the event that a new software update involves faster processing speeds. Regarding space hardware, some shifts in space architectures

---

[31] MITRE, 2014.

are likely required to enable broad enterprise-level MOSA implementation. There is interest in taking advantage of common hosted payloads, such as space weather and threat warning and response sensors, to improve space situational awareness and the resilience of the enterprise.[32] As threats advance and evolve, threat warning sensors may need to be upgraded and fielded on a more frequent basis. With a MOSA approach (e.g., establishing common connection and communications standards), sensors with the latest technology could be rapidly integrated on to a satellite just prior to deployment.

Additionally, as space systems move toward architectures with an increased number of satellites and more frequent replenishment rate,[33] designing in modularity could overcome the issue of low production volume limiting the business case for MOSA implementation. Further, if the current trend of LEO mega-constellations in the commercial space market continues,[34] there may be an opportunity for the standardization of some satellite component interfaces. AFSPC and SMC should continue to look for opportunities to leverage commercial developments. In the extreme, standardizing bus hardware could drive toward off-the-shelf buses or components that could be quickly modified or upgraded before launch. If the technical and business case can be made, this has the potential to create an environment where satellite systems could be launched on a reliable schedule, similar to A-RCI installations on board submarines.

### *Select Appropriate Interface Standards*

The selection of appropriate standards is another key aspect of up-front systems engineering. In the absence of voluntary standardization,[35] efforts must be taken to sponsor adoption or agreement on common

---

[32] SMC/AD staff, 2017; SMC Space Defense Task Force, 2017.

[33] Or, at least, satellite building blocks, like CubeSats (Charles Gustafson and Siegfried W. Janson, *Think Big, Fly Small*, The Aerospace Corporation, September 1, 2014).

[34] Thierry Dubois, "Eight Satellite Constellations Promising Internet Service From Space," *Aviation Week and Space Technology*, December 19, 2017.

[35] The Universal Serial Bus standard is an example of voluntary standardization. See Appendix A for more details.

interface standards. The use of common standards for an enterprise-level system architecture allows enterprise managers to reduce rework and take advantage of related economies of scale to achieve cost savings.[36] Enterprise design agents can champion the use of a particular standard by investing in the development of that standard, either through intensive government involvement in technical details of standard development or establishment of an industry consortium to develop the standard using industry's technical expertise. If standards use is championed at the system (program) level, rather than coordinated at the enterprise level, rework and disagreements may ensue. It is therefore necessary to have this effort performed at the enterprise, vice program, level.

The importance of championing the broad use of the selected standard cannot be overstated. Without a broad base of users, even an openly developed standard provides no additional technical or business benefit to a proprietary standard, and support for such a standard will wane as its user base decreases. However, enterprise governance should engage in decisions on which standards to invest in so as to avoid waste and conflicting approaches in the future.

OMS is an example of a standard developed through an industry consortium.[37] The OMS consortium is organized through AFRCO and is made up of aircraft and mission system design agents who actively develop and evolve a MOSA for aircraft mission systems.[38] While OMS has seen success to date, industry consortiums are not necessarily an automatically successful approach to developing and managing new interface standards. An issue that can plague industry consortiums is a tendency to shift from establishing common interface points toward a "design by committee," where no one entity can provide a ruling

---

[36] MITRE, 2014.

[37] Sledge, 2015.

[38] Lockheed Martin Corporation, *Lockheed Martin Demonstrates Continued Commitment to Air Force OMS Vision*, Palmdale, Calif., December 10, 2014.

on debate between multiple parties.[39] Such a situation may result in a bloated system with unnecessary complexity.[40]

Another issue with consortiums is that if the standard being developed is not tied to a specific commercial system, a government program for which that entity is competing, or some other incentive (e.g., funding for standards development), it is unlikely that industry will be fully engaged in a new standard development effort. This was the case in the SUMO initiative.[41] Without a long-term incentive for participation in a consortium, there is no guarantee that industry will remain engaged in that standards development effort. In a situation like the space market without a single overarching customer with dominant market pull, DoD could still influence the market by adopting a common standard for all of their own applications. Should MOSA have a role in the space segment of the SEV, AFSPC and SMC should dedicate sufficient resources and time to proactively manage standards development efforts. On the ground segment, the Air Force could affect market demand. As mentioned in Chapter 1, EGS and ESBMC2 will each use a MOSA. If interoperability for these two systems are desired in the future, making open design decisions—such as common communications protocol—would be beneficial in the near term. Driving as many DoD systems as possible to common software infrastructure and interface standards itself could provide a larger market volume to motivate commercial investment in a standard.

### Overarching Recommendation 3: Develop Up-Front Life-Cycle Business and Program Plans

In parallel with the development of a technical plan, AFSPC and SMC should consider program business engagement and programmatic plans that address issues across the life of the systems. Business engage-

[39] There may be other concerns associated with the design-by-committee approach, such as achieving mediocre outcomes.

[40] Sledge, 2015.

[41] Boeing staff, 2017; Orbital ATK staff, 2017.

ment plans include data rights strategies and selection of an integrator. Programmatic plans involve collaborative interaction on mandatory program structure questions and early development of a program life-cycle plan.

#### *Develop Business Engagement Plans to Enable Competition*

Ensuring sufficient ownership of data rights can be an important factor for MOSA to enable third-party system integration and in some cases where alternative suppliers are desired (e.g., for recompetition).[42] When developing a modular system design for the SEV, the Air Force should determine the level of knowledge it requires for the overall design and each component to achieve the benefits they seek. While purchasing full data rights for a unique component enables future production options, it is typically very expensive and may not be for sale from some component suppliers. In the event that component suppliers wish to keep portions of their designs proprietary, multiple alternatives are available to the Air Force. One potential option is the use of *enabling* data rights clauses to allow support contractors limited access to additional data rights for maintenance or integration functions.[43] A few industry partners we spoke with indicated that they were familiar with contractual stipulations to enable third-party integration and are able to work in such an environment. Similarly, the early selection of a system integration approach (i.e., choosing between a government and contractor integrator) provides an opportunity for the Air Force to engage with a contractor and tailor data rights strategies to suppliers' specific roles and needs.

#### *Develop Integrated Life-Cycle Program Plans*

Lastly, considerations for a system's life-cycle needs must be made early in system development. This includes using operational considerations to inform development, testing, fielding, and sustainment time lines.

---

[42] We note that in many cases, the standardized interfaces are sufficient to enable system integration, and in some cases, the increased demand from enterprise-wide use of a standard may draw new vendors to produce an otherwise unique component. Further, MOSA enables the competition of other components even if some remain unique and sole sourced.

[43] Data rights concerns are discussed in Appendix A.

As in the case of A-RCI, crew training needs, operational needs, and maintenance schedules should drive a given development schedule, rather than the development schedule driving a particular delivery date. A-RCI's transition to a reliable calendar-driven POR within the JCIDS and test and evaluation processes was in part related to the ability to decompose capability upgrades into hardware or software insertions that could be designed or procured according to their preset schedule. Give-and-take between life-cycle program plan development and the systems engineering effort is paramount to ensure that technology development time lines are fully supported.[44]

## Further Considerations for Air Force Space Command and Space and Missile Systems Center

Our analysis revealed that implementing MOSA successfully demands substantial up-front work that spans across such areas as architecture development, requirements determination, enterprise governance development, systems engineering, and industry engagement to ensure that a favorable environment for MOSA is created. For instance, AFSPC would need to review existing and prospective architectures to identify areas that can be modularized from technical, operational, and programmatic perspectives. Further, key activities associated with MOSA could have implications for organizational roles and responsibilities. More in-depth analysis may be needed to determine exactly how to implement such organizational changes once AFSPC and SMC identify where MOSA would be applied. Additionally, the increased up-front planning and systems engineering, in conjunction with the limited experience in MOSA at AFSPC and SMC, would likely require an increase in training on systems engineering and MOSA-related topics and could require potential increases in systems engineering capacity.

Establishing a MOSA environment may require a commitment to implement a MOSA approach well before program initiation and may

---

[44] For more discussion on A-RCI's incremental program management and technology development approach, see Appendix B.

take time to develop and mature. AFSPC and SMC should appropriately plan for these upfront activities in terms of schedule and resources. Another key implication to consider is that the enterprise perspective and incremental development/fielding associated with MOSA may initially deliver a less-than-optimal solution for any one individual program or stakeholder, though still delivering a useful capability and responding to emerging threat and technology in a timely manner. This would require the broader acquisition community (to include the requirements, user, and testing communities, for example) to view risk and "requirements" differently than how they are viewed in the traditional acquisition DoD setting. AFSPC and SMC would need to consider these trade-offs associated with the MOSA approach to determine how and where to implement MOSA appropriately if responsiveness and enterprise synchronization is a priority.

CHAPTER THREE

# Agile Acquisition

In this chapter, we describe some common challenges in implementing agile acquisition in a federal setting and potential ways to addressing these challenges as AFSPC and SMC consider expanding the use of agile acquisition in the SEV. For readers not familiar with agile methods, we recommend referring to Appendix C in the online appendixes for a general discussion of key attributes and practices of agile methods, as well as key differences between agile and traditional acquisition approaches.

## Application and Limitations of Agile Acquisition in the Space Enterprise Vision

Agile acquisition is feasible only when a development task can be broken into usable pieces and provided to users in a sequence as each piece is completed.[1] For example, an effort to develop a product with critical safety requirements or a need for a high level of precision in performance should not be delivered to users until the product achieves certain critical minimum levels of performance. Because users cannot apply the new products in a real operational setting until these thresholds are achieved, the users cannot rely on such applications of the products to provide feedback on next steps in development. Only when

[1] This characteristic makes MOSA an enabler for agile acquisition.

products can be incrementally upgraded can such interaction between users and developers occur and enable agile approaches.[2]

The incremental aspect of agile development is one reason why agile acquisition has found such favor in software development. Code can often incrementally add new features. Code can be rewritten relatively easily to refine existing features, especially when automatic testing tools can occur in the background during development.

Thus, an agile approach can potentially be applied to space software but likely would be harder to apply to hardware in space.[3] Each new launch can potentially deliver hardware with new incremental features. But hardware already present in space is much harder to alter and thus upgrade.[4] Hardware that remains on the ground, of course, such as that present in a ground control station, does not present this problem. Ground station hardware could be a potential candidate for agile acquisition.

Thus, the ground segment may be more amenable to agile acquisition approaches in the near term while AFSPC and SMC accumulate experience in employing these approaches and develop capabilities. Applying agile approaches to software development and the ground segment provides an opportunity for AFSPC and SMC to build upon that experience and later apply agile approaches to the more complex environment associated with spacecraft hardware acquisition.

## Agile Acquisition Implementation Challenges

Software developers have been trying to employ agile acquisition practices throughout the federal government and in DoD and the Air Force

---

[2] GPS is trying to do this with the *tech inject* points on its GPS III SV 11-32 acquisition strategy.

[3] Opportunities exist to enable an agile approach by substituting software for hardware as, for example, in software-defined radios.

[4] When this occurs, the Air Force might choose to launch a system with limited capabilities to get that capability on orbit earlier. Additional capabilities could be added on later launches when additional development was complete.

in particular.[5] Federal standards associated with requirements determination, acquisition rules, and budgeting systems often constrain the application of agile acquisition practices within a government setting. But the number of success stories is growing, and continuing efforts offer insights into how agile practices can be accommodated in ways that yield cost-effective outcomes.[6] These experiences also highlight a number of implementation challenges that repeatedly arise, which AFSPC and

---

[5] For useful overviews, see Government Accountability Office (GAO), *Software Development: Effective Practices and Federal Challenges in Applying Agile Methods*, Washington, D.C., GAO-12-681, 2012; Mary Ann Lapham et al., *Considerations for Using Agile in DoD Acquisition*, Pittsburgh, Pa.: Carnegie Mellon University, Software Engineering Institute, Technical Note, CMU/SEI-2010-TN-002, April 2010; Mary Ann Lapham et al., *Agile Methods: Selected DoD Management and Acquisition Concerns*, Pittsburgh, Pa.: Carnegie Mellon University, Software Engineering Institute, Technical Note CMU/SEI-2011-TN-002, October 2011; Nancy Markuson and Joleen Flasher, *Adapting Agile Processes for Military Acquisition Programs*, briefing, McLean, Va.: MITRE, April 1, 2014; Suzanne Miller et al., *Update 2016: Considerations for Using Agile in DoD Acquisition*, Pittsburgh, Pa.: Carnegie Mellon University, Software Engineering Institute, Technical Note CMU/SEI-2016-TN-001, December 2016; Peter Modigliani and Su Chang, *Defense Agile Acquisition Guide: Tailoring DoD IT Acquisition Program Structures and Processes to Rapidly Deliver Capabilities*, McLean, Va.: MITRE, March 2014; Carlton Northern et al., *Handbook for Implementing Agile in Department of Defense Information Technology Acquisition*, McLean, Va.: MITRE Corporation, Technical Report MTR100489, December 15, 2010; Amber R. Oar et al., "DoD Acquisitions Reform: Embracing and Implementing Agile USAF," *Air and Space Power Journal*, Vol. 29, No. 6, 2015; and Rex B. Reagan and David F. Rico, "Lean and Agile Acquisition and Systems Engineering: A Paradigm Whose Time Has Come," *Defense AT&L*, November–December 2010, pp. 49–52. This chapter draws heavily on these sources. Agile acquisition is increasingly finding application to hardware development, but it is most commonly found in software development.

[6] Federal experience to date has led to an analog to the agile manifesto presented in Kent Beck et al., "Manifesto for Agile Software Development," 2001, called the "FIST Manifesto." The acronym highlights the power of fast, inexpensive, simple, tiny forms of acquisition: "System development projects should be done by the smallest possible team of talented people, using a short schedule, a small budget and mature technologies to deliver innovative solutions to urgent needs. Short timelines increase agility and stabilize requirements, technology, budgets and people. Short timelines also force accountability, ownership and learning. To maintain short timelines, a project must also exercise restraint over budgets, complexity, and size. Increases to the project's budget, complexity or size inevitably reduce its speed" (Miller et al., 2016, pp. 61–62). For details on the background behind this approach, see Dan Ward, *FIRE: How Fast, Inexpensive, Restrained, and Elegant Methods Ignite Innovation*, New York: HarperBusiness, 2014.

SMC should be aware of if they choose to use agile methods.[7] We list common implementation challenges below and discuss them in turn in this section:[8]

- tailoring agile to specific circumstances and choosing the right hybrid
- determining requirements to get the full benefit of agile acquisition methods
- ensuring that developers and those who oversee them have appropriate training and experience
- maintaining timely, committed user participation.

#### Tailoring Agile Acquisition to Specific Circumstances and Choosing the Right Hybrid

Agile acquisition practices are not appropriate for every acquisition setting. In some cases, agile can yield better outcomes than traditional, plan-driven methods; in other settings, traditional methods perform better. And in still other cases, a hybrid approach that combines both methods, and is perhaps dominated by traditional methods, provides the best outcome. Early users have had difficulty knowing which approach is better for a given circumstance.[9] This is particularly true with a hybrid approach that, if not appropriately designed, could yield outcomes that are less optimal than either approach employed on its own. Users and developers have worked to develop acquisition approaches that draw the best from traditional and agile concepts, but designing hybrids that achieve is still new and, thus, remains challenging. For example, it is challenging to find a hybrid tool that enables management and oversight while not

---

[7] JMS Increment 2 initially applied agile for its acquisition and faced many of these challenges (SMC, Space Superiority Systems Directorate staff, discussion on lessons from JMS Increment 2, El Segundo, Calif., January 20, 2017). See Appendix E for our observations on the difficulties that JMS faced in employing agile.

[8] For a more complete and detailed discussion of implementation challenges observed in a federal setting, see GAO, 2012.

[9] For a useful framework for comparing traditional and agile approaches in any particular setting, see Barry Boehm and Richard Turner, *Balancing Agility and Discipline: A Guide for the Perplexed*, Boston: Addison-Wesley, 2004.

interfering with the features of agile.[10] Users and developers are still learning. Knowing what hybrid will work best in a particular setting remains a work in progress.

Getting the right mix of traditional compliance and agile tailoring requires an understanding of how applicable different federal regulations are to the agile process and how they achieve desired outcomes. Unfortunately, current DoD acquisition training tends to focus on compliance rather than helping trainees understand the importance of pursuing better outcomes. A key principle of agile acquisition is tailoring an acquisition to the specific needs of a particular circumstance and system. The personnel in a development effort can choose what specific steps to take by maintaining an agile state of mind at all times.[11] That means that applying all the controls in a traditional acquisition may not be necessary and, in fact, may be wasteful.[12] No simple checklist can tell developers how to be agile in this sense;[13] experience is ultimately the best teacher. As a practical matter, if supervisors allow them to (as they are supposed to), federal acquisition personnel learn how to achieve better outcomes on the job through accumulating experience. It has proven to be challenging for less-experienced federal

---

[10] Traditional acquisition relies heavily on management and oversight tools, such as earned-value management (EVM) systems and sometimes on detailed incentive structures linked to EVM metrics. Unfortunately, EVM is not well suited to agile acquisition. EVM builds detailed schedules of activities and resource consumption and tracks project execution against the schedules; agile assumes that details about a project schedule cannot be projected far in the future, making it difficult to build such schedules. A pure agile approach relies on demonstrated performance to date to justify each additional increment if required. If a development is no longer delivering enough value to justify its cost, the development effort ends. Federal personnel have devised hybrids of these approaches, but all tend to limit the flexibility that agile potentially allows. For example, some have applied a burn down chart as an analog to EVM tools. See Appendix C in the online appendixes for more information.

[11] For a more extensive discussion of this change in mind-set, see McKinsey Agile Tribe, *The Five Trademarks of Agile Organizations*, New York: McKinsey and Company, December 2017, p. 9.

[12] For example, documentation is limited to what is required and essential, not what is prescribed by policy or regulation.

[13] One reviewer of this report noted that more sophisticated situation-specific guidance and decisions trees could help.

personnel to appreciate how agile they can be and how being agile can improve their outcomes.

### Determining Requirements to Get the Full Benefits of Agile Acquisition

A traditional DoD approach formulates capability requirements, translates these into engineering specifications, uses these specifications to access resources through the budgeting process, and then starts a formal acquisition process designed to ensure that the developer achieves the technical specifications that were initially defined, often many years earlier. Rigidity and predictability are deliberately designed into the approach.

An agile approach explicitly rejects this strict, linear approach. In agile, the goal is to approach requirements determination as an exercise in defining the commander's intent—that is, to emphasize capability rather than engineering specification. It also seeks to identify the minimum level of performance that is acceptable to allow the operational application of a new product and to define capability requirements with enough "headroom" that developers can upgrade performance incrementally in directions that users identify will be most useful to them by the time development is complete enough to give them a usable product. Users and developers feel their way forward toward useful operational attributes for new systems. Users are still learning how to do this well.

Requirements development for SEV is focused on determining a warfighter's risk-accepted level of performance that must be accomplished on time (to outpace future threats) and within specified cost. Thus, AFSPC is currently working on identifying "warfighter essential capabilities" to develop requirements from a capability perspective, which would allow flexibility at the system level as long as the Air Force can prove that all requirements are being met at the enterprise level.[14] In this approach, the relevant commander is the Combatant Commander.

---

[14] AFSPC, "Space Mission Assurance Framework," September 2017b, Not available to the general public; AFSPC is shifting to use "essential" rather than "minimum" capability or requirement because "essential" denotes that it is what warfighters must have and is determined by them based on risk-based analysis.

### Ensuring that Developers and Those Who Oversee Them Have Appropriate Training and Experience

Government personnel regularly face challenges implementing agile approaches when they rely on developers that lack relevant experience and when they themselves are not experienced enough to get the most from their contractors. Appropriate training can alert government personnel who are new to agile acquisition to the challenges they will face. It can ensure that government personnel are familiar with the specific agile approach that a development contractor will employ. It will be most effective if it informs officials who can shape flexible capability requirements so that an agile approach has the freedom to adjust a design as information accumulates. Some joint training may help government and contractor personnel ensure that they share the same understanding of what agile acquisition is and how they will pursue it.

However, little training of this kind is available today from government sources, such as the Defense Acquisition University (DAU), the dominant trainer for the defense acquisition workforce.[15] DAU is working on adding information on agile acquisition to its course offerings,[16] but relying on DAU too heavily for training will be problematic for multiple reasons. First, plan-driven acquisition practices dominate DAU training, even though best practice in commercial companies is moving more and more toward agile practices, especially in software development. Second, DAU's standard approach to training emphasizes compliance. Agile acquisition is inherently more about best practice than compliance. Finally, agile acquisition is a bundle of best practices that is evolving relatively quickly. It does not lend itself to the type of certification courses at DAU unless they are kept current and are augmented by continuous learning. The Air Force may need to look to nontraditional sources in academia and the commercial sector for appropriate training. For instance, the case-based training that domi-

---

[15] The only DAU course mentioning "agile" in the 2017 catalog is ISA 320 (Advanced Program Information Systems Acquisition). This course is only four and a half days long and covers not only agile but a wide range of software topics for senior personnel.

[16] Miller et al., 2016.

nates academic business schools could teach trainees what they can do despite apparent regulatory constraints.[17]

Other government personnel who must engage with agile acquisition, requirements developers and users, also require experience and skills to make it work well.[18] However, DAU focuses its training on the roles of personnel in the defense acquisition workforce, not requirements developers and users. Without their participation, effective training for the acquisition workforce based on role playing is incomplete. And in many ways, agile works best when they change their roles even more than the personnel in the acquisition workforce do.

Further, personnel who write requirements or work in operational user positions typically rotate in and out of these positions, making it hard for them to gain useful experience or a long-term relationship with others involved in agile acquisition. Without effective training, it will be difficult to expect them to participate effectively. And without their effective participation, agile acquisition is limited in the value it can add. Without appropriate requirements, much of the flexibility that agile can support evaporates. Without effective user participation, developers get ineffective feedback on what users want and on how to orient any particular acquisition as it proceeds.

### Maintaining Timely, Committed User Participation

Users—or representatives of their interests as stakeholders—participate much more in an agile acquisition than they would if the system was acquired in a traditional process.[19] Users actively participate to frame acceptance tests at the close of each development increment. Working closely with contracting specialists, they determine how to direct each additional increment. They play an active role in deciding when additional iterations are no longer needed. And in the most aggressive forms of agile acquisition, they are present and accessible to developers

---

[17] One reviewer of the report noted that a training program that also explores and explains the models and theories of acquisition approaches would be helpful.

[18] See Appendix C for more details on how user needs are captured in agile acquisition.

[19] See Appendix C for additional detail on users' role in agile.

on a daily basis through the course of each development increment. If users do not effectively participate, the benefits of agile acquisition cannot be fully realized.[20] One of the challenges in obtaining user participation is that government organizations must commit the time of these users, typically by removing other commitments. Additionally, there needs to be continuity in the users engaged in development over a series of increments. Both aspects are challenging. Another key challenge is that the government must choose which users or user representatives to involve in development activities. This can be particularly challenging for space acquisition because space-based systems often provide services to diverse users well removed from the Air Force. In this setting, AFSPC may need to choose to provide surrogates for these users.

## Overcoming Challenges to Implementing Agile Acquisition

If AFSPC and SMC decide to use agile acquisition in support of the SEV, they will need to overcome the challenges described above. Toward this end, our analysis identified potential ways to overcome these challenges to improve the likelihood of the successful implementation of agile in the SEV. We summarize our findings as three overarching recommended actions:

- build and sustain an experienced workforce for agile acquisition
- create an enabling environment for agile
- build "agile" into acquisition strategy.

Table 3.1 lists specific suggested actions associated with each of these, indicating which ones apply to AFSPC or SMC. The following subsections explain each action in more detail, synthesizing findings

---

[20] See, for example, GAO, 2012; Lapham et al., 2010; Lapham et al., 2011; Miller et al., 2016.

**Table 3.1**
**Recommendation Actions for AFSPC and SMC to Enable Agile Acquisition**

| Recommendation | AFSPC | SMC |
|---|---|---|
| *Overarching Recommendation 1: Build and Sustain an Experienced Workforce for Agile* | | |
| Build a workforce with appropriate skills and continuity working together | X | X |
| Build support of the senior Air Force leadership | X | X |
| Make agile experience a prerequisite for advancement in areas where you expect to use agile in the future | X | X |
| Maintain regular contact with others conducting agile to stay abreast of best practices | X | X |
| *Overarching Recommendation 2: Create an Enabling Environment for Agile* | | |
| State requirements to allow ongoing flexibility as relevant threats and technologies change | X | |
| Work with relevant user communities to ensure effective user participation | X | X |
| Work with the contracting community to create a seamless way for users to direct developer activities in compliance with Federal Acquisition Regulation (FAR) | | X |
| Work with the test community to certify and coordinate internal and external tests focused on realized operational capability | X | X |
| *Overarching Recommendation 3: Build "Agile" into Acquisition Strategy* | | |
| Frame the request for proposal (RFP) and contract materials to facilitate agile | | Mainly |
| Build a calendar-driven battle rhythm that assesses status at fixed intervals | | X |
| When possible, rely on trust, reputation to ensure delivery of what is promised to limit the need for active oversight | | Mainly |

from our analysis of interviews and the application of agile acquisition to DoD from a range of guidebooks.[21]

[21] See Lapham et al., 2010; Lapham et al., 2011; Markuson and Flasher, 2014; Miller et al., 2016; Modigliani and Chang, 2014; and Northern et al., 2010.

## Overarching Recommendation 1: Build and Sustain an Experienced Workforce for Agile Acquisition

Agile acquisition works best when the people involved have significant experience with it, have worked with one another on agile acquisition in the past, and have sufficient skills to make appropriate decisions without detailed instructions. These qualifications apply to developers, those who work with them, and those who oversee all elements of the acquisition from requirements determination to software sustainment. Today, relevant skills are growing on all fronts. Because agile acquisition evolves over time, skills must continue to grow on all fronts for the participants to stay current.

AFSPC and SMC are likely to need a formal workforce-shaping program to do this. Such a program should seek to learn from realized experience to build improving performance over time. Below we discuss a set of actions that could be part of such a workforce-shaping program.

### *Build a Workforce with Appropriate Skills and Continuity Working Together*

Although skills in agile are beginning to accumulate in acquisition activities within SMC and other closely related activities, such as the AFRL, it will take time for SMC to accumulate experience unless it hires laterally from other organizations already engaged in agile acquisition.

SMC can accelerate training its staff through rotations at other government or industry organizations with more agile experience.[22] Additionally, SMC can engage external experts that specialize in supporting the implementation of agile acquisition.[23] For example, many federally funded research and development centers (FFRDCs), readily accessible to AFSPC and SMC, maintain growing practices in this kind of expertise.[24] They can provide on-site support to personnel

---

[22] This is happening today in the BMC2 JEON and AFRCO's prototype work (see Chapter 5). SMC should seek other opportunities to exchange personnel with organizations where agile acquisition is under way, even if that acquisition is not directly related to ESBMC2.

[23] GAO, 2012.

[24] Our references include multiple guides from FFRDCs, such as Software Engineering Institute (SEI), MITRE, and the Aerospace Corporation. See, for example, Lapham et al., 2010, and Modigliani and Chang, 2014.

who will participate in an agile acquisition when independence from commercial interests is needed. Commercial trainers are also available. Additionally, agile coaches can work hand in hand with government personnel involved in an agile acquisition to help them understand the nuances of the agile approach.[25] SMC can also enroll personnel in classroom training in-house or from third parties. Training is likely to work best if it occurs shortly before a particular acquisition and allows role-playing with others who will participate in the acquisition. Because the essence of agile acquisition is tailoring, the training should focus less on compliance than on problem-solving to find creative ways to use current authorities to get what users need most.[26] Ideally, coaches could tailor training to the upcoming acquisition and then walk trainees through the specific tasks they will face. This approach will prepare the coaches and trainees for their roles when the real acquisition occurs. To be most effective, this training should reflect the forms of agile acquisition that the developers will actually use during the acquisition.[27]

AFSPC and SMC can further improve long-term training in agile acquisition by seeking nontraditional sources for it. Some can come from non-DoD sources that have had more experience with agile acquisition. Some can come from internal sources. AFSPC and SMC could also create a Center for Lessons Learned that captures the experi-

---

[25] William Broadus, "The Challenges of Being Agile in DoD," *Defense AT&L*, January–February 2013, pp. 5–9, suggests using them to identify the risks that each set of stakeholders face in an agile acquisition and actions they can take to mitigate these risks.

[26] Frank J. Anderson, Jr., *A Plan to Accelerate the Transition to Performance-Based Services: Report of the 912(c) Study Group for Review of the Acquisition Training, Processes, and Tools for Services Contracts*, Washington, D.C.: U.S. Department of Defense, Report AF903T1, 1999, pp. 50–75, presents a plan for training on performance-based acquisition. Its approach would apply here as well. For examples of how USSOCOM accelerates acquisition without using any special authorities or waivers to the FAR, see Tim Troup, "Acquisition Truths from the Trenches," *Defense AT&L*, May–June 2010, pp. 42–45. Cf. Miller et al., 2016, p. 14.

[27] GAO, 2012; Miller et al., 2016, p. 43. This is challenging to do today when the specifics of agile acquisition are often in flux during an acquisition. It should become easier as preferred standards emerge in a federal setting.

ences from personnel involved in agile acquisition and have such information easily accessible for less-experienced personnel.[28]

### *Build the Support of the Senior Air Force Leadership*

Agile development works best when all of its elements and its enablers are aligned to leadership priorities, from end to end—requirements determination, programming, acquisition (from strategy through contracting to program management and testing), and user participation.

Two types of specialists can help obtain and sustain leadership engagement and support. One is an agile champion in AFSPC that could take on full-time responsibility to monitor all the moving parts of agile acquisition.[29] Ideally, this person would be an O-6 with broad experience and name recognition across the requirements, programming, acquisition, and operational communities involved in the implementation of agile acquisition. The champion's primary responsibility would be to translate information on agile implementation into operational terms relevant to the leadership and use that to advise the leadership on what the implementers need from the leadership to sustain their effort.

A second specialist is an agile advocate[30] who can advise policymakers at relevant echelons within AFSPC and SMC on what they can do to ensure that an agile acquisition runs effectively.[31] They need the respect of the policymakers they are advising, but they need not

---

[28] Commercial companies in dynamic markets routinely rely on their own personnel to teach what they have learned in recent assignments to other personnel. The U.S. Army also experienced something similar when young leaders discovered in recent deployments that their formal training did not prepare them for the battlefields they experienced. To keep up with the dynamic nature of the conflicts they faced, they increasingly turned to such sources of information as the Center for Army Lessons Learned, where they could find timely advice from their peers that was more relevant to their needs than the formal courses painstakingly developed at home.

[29] GAO, 2012. The agile champion would chair a working group of all the stakeholders involved in agile acquisition. With the working group, the champion would gather information on the status of agile acquisition relevant to AFSPC, report that to a senior, flag-level advisory board representing the same stakeholders, take guidance from the advisory board, and communicate that to all personnel involved in the implementation of agile acquisition. For details on the working group, see Appendix E in the online appendixes.

[30] *Agile advocate* is a term of art for a trainer who serves during a specific development as an on-site mentor for the personnel actually executing the development.

[31] Broadus, 2013, pp. 8–9.

be government employees, and they need not be assigned full-time in one development program. In particular, they need enough skills in modern agile software development and associated acquisition practices to be persuasive.[32]

Additional external experts could be available on a regular basis to field questions and to offer proactive advice. To be most effective, they should coordinate their work with the agile champion and align their advisory services with the champion's communication plan.[33]

#### *Make Agile Acquisition Experience a Prerequisite for Advancement in Areas Where You Expect to Use Agile Acquisition in the Future*

Agile acquisition for the most part is a body of practices derived from Lean Six Sigma methods. What has worked for Lean Six Sigma could give indications of what might work for agile acquisition. A best commercial practice for Lean Six Sigma is making Lean Six Sigma a core business skill for managers rather than maintaining it among specialists who operate outside the core product lines of a business. To make agile a core skill, AFSPC and SMC (and other Air Force organizations that prepare personnel to serve in AFSPC and SMC) might consider giving their line commanders and directors immediate personnel experience with agile acquisition. As an example, General Electric, well known for its effective development of leaders, does not allow anyone to become a general manager unless he or she directly supervised a successful Lean Six Sigma initiative. An analogous policy would not allow an O-6 or GS-15 to lead a software development or acquisition effort without having led a successful agile development or acquisition effort earlier. This would make agile acquisition a core skill in SMC rather than a form of expertise maintained beyond the chain of command in a technical support organization.

#### *Maintain Regular Contact with Others Conducting Agile Acquisition to Stay Abreast of Best Practices*

As experience accumulates, best practices morph, as they do in all ongoing efforts to improve processes and performance. So we should

---

[32] Miller et al., 2016, p. 43.

[33] Appendix E provides more information on this topic.

expect the best practices of agile acquisition to continue to change.[34] And as needs continue to evolve, software development practice will as well.

In dynamic environments, Lean Six Sigma teaches organizations to maintain an active engagement with the external environment. Participation in formal process benchmarking circles, trade meetings, and other professional networks all offer places to learn. Benchmarking is an especially good way to stay current with best practices in agile as they evolve because commercial companies in very different industries can use similar agile practices. An organization benchmarks itself against other organizations doing similar things by learning from public sources, then signing agreements to share sensitive information, and finally by imbedding personnel in another organization so that they can access latent knowledge—knowledge that is difficult to articulate and write down—on practices they would like to emulate.[35] Such latent knowledge is especially helpful for agile acquisition, because it contains elements of practice that can only be learned through direct experience. Because companies in different industries do not compete, they are willing to share information and in fact seek opportunities to learn from others to stay as current as possible as agile acquisition continues to advance.

### Overarching Recommendation 2: Create an Enabling Environment for Agile Acquisition

Early critics of agile acquisition thought it lacked the discipline of plan-driven acquisition. In fact, agile acquisition requires close discipline to

---

[34] For example, as the cycle times for updating software continue to shrink, a new approach to developing software called DevOps (for development and operations) is emerging from agile practice.

[35] For more information on these approaches to benchmarking, see Frank Camm, "Adapting Best Commercial Practices to Defense," in Stuart E. Johnson, Martin C. Libicki, and Gregory F. Treverton, eds., *New Challenges, New Tools for Defense Decisionmaking*, Santa Monica, Calif.: RAND Corporation, MR-1576-RC, 2003, pp. 211–246. Latent knowledge in this setting is practical working knowledge about a particular activity that is conveyed through hands-on practice and is typically not captured in technical orders or other forms of formal documentation.

create an environment that is safe for rapid learning and creativity.[36] Advance planning is required to prepare such an environment. Agile acquisition takes the time to do this planning because it allows much more rapid development once development actually begins. Much of this planning involves ensuring that government and contractor development teams have the experience, skills, and tools to work well together with limited top-down oversight.[37]

#### *State Requirements to Allow Ongoing Flexibility as Relevant Threats, Technologies Change*

Agile acquisition can achieve most of its goals with a requirements document that has two attributes. First, it states the requirement in terms of a capability to execute a mission at an acceptable level of risk, not in terms of one material engineering solution that might be employed to execute a mission.[38] If the mission is stated broadly enough, an agile acquisition can adjust its focus as circumstances surrounding that mission change without changing the mission itself.

Second, it states the minimum requirement that a user will find valuable.[39] Once this minimum is defined, an agile development can effectively use it as the goal to achieve in the first release to

---

[36] Tricia Oberndorf et al., "Finding Discipline in an Agile Acquisition Process," briefing presented at the 23rd Systems and Software Technology Conference (SSTC), May 16–19, 2011, Salt Lake City, Utah, Carnegie Mellon University, Software Engineering Institute, 2011, p. 19.

[37] In effect, agile acquisition applies something like mission-type tactics that employ well-prepared personnel and tools to execute the commander's intent stated in terms of mission goals. A practice that began in the military migrated into best commercial practice and is now finding its way back into DoD (Broadus, 2013, pp. 8–9). For a description of this approach, which the Germans who developed it initially called *Auftragstaktik*, see David M. Keithly and Stephen P. Ferris, "Auftragstaktik, or Directive Control, in Joint and Combined Operations," *Parameters*, Autumn 1999, Vol. XXIX, No. 3, pp. 118–133.

[38] As discussed earlier, AFSPC is currently working toward such a goal with the use of "warfighter essential capability" (AFSPC, 2017b).

[39] This is not equivalent to a threshold as it is currently used in an acquisition. It is common for a threshold to reflect everything users believe they will ultimately need, because they fear they will get only one chance to ask for the money to meet their needs. It is risky to rely on future decisions to allow realized capability to expand over time.

users.[40] Agile acquisition normally defines a release in terms of a date; the release delivers whatever capability developers can provide by that date. In effect, the first release in an agile acquisition must be capability driven rather than calendar driven if developers cannot achieve the minimum capability by the scheduled release date. Once developers achieve the minimum, additional increments can be defined to reflect the priorities that prevail and the resources available for each increment.

The current version of the JCIDS process promotes a requirement of just this form for significant software initiatives that do not require associated hardware development.[41] It requires joint review of only an initial capability document (ICD) no more than ten pages long. "Capability requirements should be general enough so as not to prejudice decisions in favor of a particular capability solution, but specific enough to evaluate alternative approaches to achieve the capability."[42] Subsequent review of details on how to proceed occurs within a military service. Otherwise, a biennial joint review tracks the pursuit of the requirement to verify that it remains within the terms defined in the ICD. In particular, key performance parameters and key system attributes can change without any joint review. In effect, the Air Force has the authority to define how flexible it wants its requirement management process to be in software development.

The resource programming process can complicate an agile approach that uses a broad mission-based definition of a requirement with a narrow minimum level of mission capability. In the competition for funds, program offices often worry that any admission of a minimum may lead to funding that covers only the minimum level. For

---

[40] Again, the "minimum" requirement in this context is for the purpose of guiding an agile development. It is not equivalent to the "essential" requirement that AFSPC is developing. AFSPC defines the essential requirement as the capability a user must have to enable success with acceptable risk in a contested environment. A greater capability will be valuable but is not essential or potentially technically feasible or affordable in required time frames to outpace threats. Thus, in an agile development, the essential requirement could be achieved in a later release, not necessarily in the first release.

[41] Chairman of the Joint Chiefs of Staff, *Manual for the Operation of the Joint Capabilities Integration and Development System (JCIDS)*, February 12, 2015, including errata as of December 18, 2015, Enclosure D.

[42] Chairman of the Joint Chiefs of Staff, 2015, p. D-25.

agile acquisition to work well, the minimum level cannot be interpreted as a "good enough" level. Participants in the resource programming process must understand that a typical development will start small—at least at the minimum level of capacity—and add capability as long as additional incremental releases add enough mission-related value to justify their incremental cost within the requirement, however formally defined.

### *Work with Relevant User Communities to Ensure Effective User Participation*

AFSPC and SMC could consider alternatives to address the challenges associated with obtaining users' (or user representatives') time commitment and participation for the duration of a development. For instance, AFSPC and SMC could establish a team of members who are expert in user needs to support developers at the end of each release and, as needed, between releases. These members need not be from a user community to represent it. A variant in particularly dynamic environments maintains a group of test specialists in a user organization who are familiar with agile acquisition and with the activities in that organization. This balances the desirability of being close to the actual user environment with the desirability of releasing user representatives from other duties.

While agile acquisition is designed to make user priorities paramount and to drive an acquisition toward their priorities,[43] multiple interests—from resource programmers, sustainment specialists, and others—also have legitimate roles in the development. Commercial best practice uses cross-functional teams to represent all of these interests in decisions. It empowers them to make quick decisions without any further input from higher-level authorities in individual stovepipes. AFSPC might retain the right to make decisions in such teams, but effective user representatives must be the first among equals to implement the intent behind agile acquisition; any team lead must have a persuasive reason for overriding a participant representing users. Given this central role, it should be clear that representing users is an inherently governmental responsibility.

---

[43] Broadus, 2013, pp. 8–9.

### *Work with the Contracting Community to Create a Seamless Way for Users to Direct Developer Activities in Compliance with the Federal Acquisition Regulation*

Agile acquisition requires arrangements in which the development rarely has to wait for the approval of changes in scope as it proceeds. This often involves placing contracting officers (COs) directly with the users who assess incremental releases and ensuring that they will be ready to accept proposed changes as soon as they occur. As such, advance arrangements must be negotiated with the COs who support the development through its lifetime.[44]

We offer two related approaches for SMC's consideration. One asks a CO to give attention to Air Force–wide goals relevant to an acquisition, as well as compliance with stated contracting policies. A knowledgeable CO can help a program comply in a way that best advances the Air Force's operational priorities. In the past, savvy acquisition managers have promoted this approach by cultivating close relationships with their COs to make them as aware of the acquisition team's priorities as they are of their professional community's priorities. When done well, this approach has led COs to learn more about what compliance really requires and to tailor arrangements to promote the acquisition team's goals.[45]

The second approach assumes success with the first. After a CO has learned to support an acquisition creatively, acquisition managers can work with her or him to align the value chain they are both part of.

[44] Under the FAR, only a warranted CO can change the formal terms of a contract. By limiting the ability of non-COs to direct contractors, the FAR seeks to ensure that the federal government obligates to pay only for things properly specified in a contract. Agile acquisition actually seeks to use frequent changes to limit such negative cost and schedule effects by catching errors when they are easy to reverse and seeking smaller changes that enable a program to find its way toward an outcome without taking unnecessary excursions that require large changes to reverse. This intent does not exempt agile acquisition from a warranted CO's authority. Agile acquisitions must work with COs to accommodate the priorities of these acquisitions.

[45] AFRCO staff, 2016; Space and Missile Systems Center, Launch Enterprise Directorate staff, discussion on prototyping and Other Transaction Authority, El Segundo, Calif., April 18, 2017; OTI staff, 2017b. While the contracting culture in DoD often discourages risk-taking associated with doing something new and untested, COs can be more creative when they are informed of the desirability to seeking better solutions and empowered within their chain of command to pursue them.

This can happen in many ways. The most visible may be a willingness to cut response times by building more slack into the CO's schedule so that the CO can stand ready for requests.[46] This, of course, requires support from the broader contracting community. The fixed calendar time stamps that define agile sprints and releases make this easier. A CO can know that, on a certain date, a user representative will need support and plan around this date.

Such coordination requires trust and mutual respect, which should build as specific acquisition and contracting personnel gain more mutual experience working together.

#### *Work with the Test Community to Certify, Coordinate Internal, External Tests Focused on Realized Operational Capability*

Testing is one of the key acquisition activities where agile and traditional approaches are highly different. Testing in agile development occurs continuously through the course of an agile sprint period,[47] and developers bring their own in-process test protocols.[48] Such key dif-

---

[46] One reviewer noted that this approach may be difficult to implement given the budget realities and already overworked contracting workforce. However, the decisionmakers would need to make such trade-offs between speed and cost.

[47] In line with lean production protocols, agile acquisition seeks to build quality into a product by identifying errors as soon as possible and correcting them before additional work proceeds based on these errors. This approach appears to slow development in the moment by diverting attention from the development process itself. But it ultimately saves time and costs by reducing the time required for rework when development is complete (GAO, 2012).

[48] These include automated testing that continuously verifies that new code is consistent with specifications. They also include periodic manual tests. When sprints are complete, developers perform additional testing and integration before a release to users. (As an aside, when new code operates beyond the bounds of the "as-is" system architecture or information assurance framework, users test new code in a "compliant sandbox," which provides an instrumented test bed without extensive technical training or risks to operations [GAO, 2012]). All of this quality control occurs before the government has to apply any quality assurance through its own test activities. Best commercial practice now recognizes such quality control efforts in a provider organization and adjusts the user organization's quality assurance process to reflect it. By certifying a provider's quality control process, the user can in appropriate circumstances accept output from the provider without inspection. More commonly, some quality assurance remains, but it is tailored to the quality control process it follows. For a discussion in the context of inspections in the aviation industry, see Frank Camm et al., *Charting the Course for a New Air Force Inspection System*, Santa Monica, Calif.: RAND Corporation, TR-1291-AF, 2013.

ferences highlight the need for adjusting how SMC frames the testing approach to ensure agile methods are employed effectively.

For one, SMC should assess a potential agile developer's test processes as part of any source selection. SMC could require the use of specific test protocols or software. It could require detailed information on test protocols and software that the Air Force test community could then use to tailor its own validation and verification tests. A common practice in lean production is to certify a source's testing protocols so that acceptance testing is not required following release. A source selection may recognize some sources as having certified test protocols. But SMC cannot delegate final integration testing or configuration control, especially when multiple developers contribute software to a single system development.[49] The RFP must state explicitly how formal Air Force testing will relate to this in-process testing. SMC could work with the Air Force test community to determine what test protocols and software SMC should require developers to include in an offer and perhaps in evaluation of offers.[50]

Program offices may need to work with the test community to conduct any quality assurance tests required at the end of a release with as little delay in calendar time as possible.[51] This approach will require adjustments in the test community that could appear to degrade its performance; today, testers can place program offices in a queue and service them to optimize the efficiency of test facilities. The clear time stamps associated with agile releases can facilitate coordination between SMC and the test community. To reduce calendar delays as much as possible, SMC should be prepared to pay Air Force testers a premium if necessary to reflect any higher costs that result from standing ready to receive an agile release for immediate testing.

---

[49] GAO, 2012.

[50] Broadus, 2013, p. 9.

[51] Another approach is for the test community to work in parallel during development and in tandem with the contractor's testing.

## Overarching Recommendation 3: Build "Agile" into Acquisition Strategy

### *Frame the Request for Proposal and Contract Materials to Facilitate Agile Acquisition*

The RFP frames the type of contract that SMC will use to access developers. Contracts should be flexible and should incentivize behavior likely to facilitate effective agile acquisition. A flexible contract can maintain a scope broad enough to cover the full range of activities relevant to the requirement for the development. It allows simple, rapid adjustments in content and level of effort within this range. A cost-based indefinite delivery/indefinite quantity contract is well suited to these needs. Formal options for individual releases can enable an incentive similar to an award term by providing a simple way to continue development as long as the developer meets SMC's expectations, broadly defined. Incentive fees can induce the behavior desired if SMC can clearly define it. For example, SMC can potentially link such fees to cost, speed with which errors are detected and corrected, percentage of backlog burned down in a release, and so on.[52] But such incentive fees often drive unintended behavior. Award fees can be extremely helpful in mitigating this risk if they are available.[53]

Any source selection needs to ensure that the individuals working for a contractor will bring appropriate experience. Thus, the RFP must state how SMC will assess the experience and skills of a development team. Effective agile development teams tend to have a higher level of expertise and are more experienced as a cohesive group than development teams used to produce similar outputs in a plan-driven setting.[54]

---

[52] Appendix C provides more information on how these things might be done. One reviewer of this report noted that cost-plus-award-fee contract (CPAF), however, does carry its own issues.

[53] For more detail, see Office of Defense Procurement and Acquisition Policy, *Guidance on Using Incentive and Other Contract Types*, memorandum, Washington, D.C., April 1, 2016.

[54] Agile skills are typically the highest among developers who have pursued agile acquisition aggressively and accumulated experience with it. Those developers need not be the ones SMC has typically worked most extensively with in the past. In fact, to the extent that Air Force policies have impeded full application of agile acquisition in the past, traditional service providers may have less experience than many nontraditional sources.

The most cost-effective team is unlikely to be the lowest cost team on a cost-per-hour basis.[55] The RFP needs to explain clearly how SMC will balance skills, experience, and cost to match its priorities. Experience and skills relevant to agile acquisition and the substance of the development work to be performed are both important. The experience and skills of the team and the program manager and other key personnel are important. Past experience in recent, similar developments can offer useful information. The RFP should detail what past performance information is important and how SMC will access and assess it.[56] This may be sufficient for building an effective agile industrial base in the long run, but SMC could be more proactive by helping to organize industry activities that enhance the agile capabilities of many sources. These could include training programs, certification programs for specific skills, and industry forums to identify the agile capabilities that SMC seeks.

Additionally, when the RFP is used to solicit offers from potential developers, it should explicitly address the following developer oversight issues to facilitate agile acquisition.

- *Coordination of scaling frameworks between Air Force and developers.* Any significant effort overseen by SMC is likely to require coordination of multiple teams, some from different developers and at different locations.[57] Scaling frameworks define the terms of reference for such coordination. The RFP must address this problem directly and identify a way forward, potentially including direction to use specific scaling tools chosen by SMC.[58]

---

[55] That does not necessarily mean that an agile approach would increase total system cost, even in a best-value competition. It is likely that a proposal to achieve some stated level of robustness is likely to have a higher price with a traditional, plan-based development approach than with an agile development approach. It is plausible that an offeror could not even achieve the requirements stated in a RFP for a system that must be robust enough to operate in a contested space without an agile approach.

[56] GAO, 2012; Miller et al., 2016, p. 40.

[57] Agile acquisition initially occurred in small teams, all of whose members were collocated. It relied heavily on the social dynamics internal to such small teams.

[58] Miller et al., 2016, pp. 24–25.

- *Protocols to manage refactoring.* Developers use refactoring to revisit completed code in a module of software and rewrite it to improve its intramodular performance in some way without changing information flows to or from the module or the role of the module in the broader system. When many teams work in parallel, protocols must be in place to ensure that refactoring by one team does not adversely affect the code written by another team.[59] The RFP must address this problem directly and identify a way forward, particularly when several different developers are working in parallel.
- *Assessment of progress.* As discussed earlier, agile acquisition requires some alternative approach to EVM that does not require a detailed description of the way forward. The RFP must define this alternative clearly.
- *Coordination of formal Air Force testing and in-process testing.* As discussed earlier, the RFP used to solicit offers from potential developers must state explicitly how formal Air Force testing will relate to developers' in-process testing.[60]
- *Acceptable versions of agile acquisition and tools.* A development team typically brings its own version of agile acquisition and tools to implement it. SMC must be clear which versions and tools it is prepared to oversee. SMC can potentially expand the capabilities available to it by having the experience and skills necessary to work with a variety of forms of agile acquisition and tools. The RFP must provide a mechanism to verify that SMC and any potential developer understands SMC's expectations clearly in the same terms.[61]

---

[59] A proper modular architecture for parallel efforts can limit such effects but not prevent them altogether. Appropriate skills and protocols present in a development team can help predict when intramodular refactoring is safe. See Miller et al., 2016, p. 26.

[60] There may be some unique challenges associated with space hardware testing. For instance, operational testing for satellites is difficult because of the remote aspect. Further investigation is needed to understand how the testing aspect would work for space hardware.

[61] Miller et al., 2016, p. 40.

### ***Build a Calendar-Driven Battle Rhythm That Assesses Status at Fixed Intervals***

Agile acquisition typically proceeds through sprints and releases that occur during fixed periods of calendar time, completing whatever work can be finished during a planned period. During each period, work proceeds at a steady rate designed to keep developers fully employed without imposing any overtime. Developers tend to design their own sprints and releases. But SMC must understand what to expect when it engages a developer. And it must ensure that all developers are effectively integrated—by SMC or a lead integrator—in the broader system program that delivers each release to users. At a minimum, all developers and SMC must have common expectations about how their mutual battle rhythm(s) will proceed. If SMC has agreements with users, COs, and external testers, it must ensure that its developers can support these agreements.

If an acquisition uses continuous integration, SMC must play a more direct role in managing the battle rhythm.

> Continuous integration may well be the most useful and controversial practice advocated in the Agile community. . . . Continuous integration of software is contingent upon the ability to concurrently execute two crucial activities: (1) collect incremental changes from multiple developers on a regular basis, ideally daily, and (2) perform the "nightly build" discipline, where all changes are brought together in an incremental software baseline, which is in turn compiled and tested with the available unit and regression tests.[62]

At a minimum, SMC must anticipate such close coordination during a source selection, ensure that all developers have the skills and experience to do this together, and assign and enforce appropriate roles and responsibilities.

---

[62] Miller et al., 2016, p. 41.

### *When Possible, Rely on Trust, Reputation to Ensure Delivery of What Is Promised to Limit the Need for Active Oversight*

To the extent that such a thing exists, pure agile acquisition highlights a relationship between developers and users. As long as users continue to be pleased with the performance of developers, a program continues. In effect, the Air Force would monitor progress in a development by assessing the degree to which a developer continues to maintain a reputation for high-quality performance.[63] Reputation would reflect the satisfaction of users and the advocates of other requirements discussed above.

In best commercial practice on services acquisition, a provider is judged and rewarded primarily on the basis of a subjective assessment of customer satisfaction. But even when this occurs, the customer and provider collect extensive data that they use to inform an ongoing discussion of the direction of effort. Standard tools used by agile developers generate metrics they need internally to control a development. In best commercial practice, a provider and user often rely on the same provider-managed system to generate metrics that interest them. This is occurring in some agile developments in the federal government.[64] Such arrangements must be resolved early in a development and potentially in the source selection.

The program office can collect many metrics without attaching them in formal incentives in the contract. SMC could use a wide variety of metrics to facilitate discussions with a developer while relying on

---

[63] GAO, 2012.

[64] Miller et al., 2016, p. 20; For instance, story points associated with user stories could be used to assess how much work is being completed during each sprint or release and how much remains in the project backlog as it is currently defined. Story points associated with classes of user stories could be used to track what portion of the requirements associated with each class has been serviced. (Appendix C describes story points and how an agile development can use them.) Such a measure would be helpful in development efforts that generate output relevant to many discrete capabilities and users. However, such measures cannot be interpreted in the same way that their analogs in an EVM system would be interpreted, because agile acquisition has no precharted plan on which to base a formal EVM system (GAO, 2012).

a small number for formal motivation.[65] As suggested above, award fees and award terms can potentially provide useful motivation for software development if SMC can make the case for their use.[66]

If mutual trust cannot be sustained in a program, it will be difficult to sustain an agile approach. When trust fails, closer oversight is usually necessary to assure quality. It is common in commercial settings to reduce oversight as trust grows and vice versa. Such a pattern could potentially yield a hybrid version of development, but it would always require a back-up plan for oversight, which could be cumbersome and could reduce developers' willingness to take appropriate risks.

## Further Considerations for Air Force Space Command and Space and Missile Systems Center

Lessons learned from DoD (and broader federal settings) on agile acquisition indicate that users have difficulty tailoring it to specific circumstances, focusing more on user priorities than on the traditional compliance concerns of the federal acquisition community, defining requirements in a way that preserves flexibility, getting an experienced workforce and training new participants, engaging users during development, and managing a development without a precise plan of execution looking forward. Overcoming these challenges will require senior leadership support, investments in human capital, and

---

[65] For more detail on how this might work, see Laura Baldwin, Frank Camm, and Nancy Young Moore, *Strategic Sourcing: Measuring and Managing Performance*, Santa Monica, Calif.: RAND Corporation, DB-287-AF, 2000; Frank Camm, "Federal Agencies Can Adapt Best Commercial Practice to Improve Their Acquisition of Services," testimony presented to the Acquisition Advisory Panel, Santa Monica, Calif.: RAND Corporation, CT-261, 2006; and Brian M. Stecher et al., *Toward a Culture of Consequences: Performance-Based Accountability Systems for Public Services*, Santa Monica, Calif.: RAND Corporation, MG-1019-EDU, 2010.

[66] On one hand, CPAF provides one practical and direct way to introduce customer satisfaction to DoD contracts and is useful in focusing a developer on the buyer's immediate priorities, which change through a development. On the other hand, CPAF may make it hard for the contractor to align their efforts to meet the incentives because the awards are subjective to the committee and change frequently.

advance planning. As AFSPC and SMC consider the application of agile acquisition in SEV, they should

- sustain senior leadership support to build and support an appropriate Air Force infrastructure based on new attitudes, experience, skills, and tools
- learn how to execute effective end-to-end, cross-functional coordination in each new agile acquisition to engage those who define requirements, allocate resources, acquire and test a new system, manage acquisition contracts, and ultimately use the new system (with effective coordination overcoming stovepipes within AFSPC, across the Air Force, across DoD, and across the user community for the system)
- seek to improve it over time, build and sustain an experienced workforce, and remain aware of ongoing trends in best agile acquisition practices, inside and outside the Air Force.

A formal change management framework can help AFSPC and SMC coordinate the moving parts involved with the above activities in a coherent manner. Further, implementing agile acquisition effectively would require a change in how risk has traditionally been viewed in space acquisition. Because agile acquisition delivers capability in an incremental way, the early increments will not be a complete or an optimal solution, though a useful solution. In agile, delivering a useful solution too late to address an emerging threat or to exploit an emerging technology is viewed as a greater risk than getting a complete solution right the first time.

CHAPTER FOUR

# Rapid Prototyping

As discussed in Chapter 1, AFSPC is currently developing and implementing a process to nominate candidates for rapid prototyping. Based on sponsor guidance, our examination of rapid prototyping focuses on the *transition* from prototype demonstration and evaluation to production and operational deployment. Our research examined potential risks associated with rapid prototyping and potential management processes and techniques that could improve the likelihood of successful transition of a rapid prototyping activity to an effective warfighting capability. This chapter summarizes the challenges of operationalizing rapid prototyping in a DoD setting and provides recommendations for AFSPC and SMC on how they might overcome these challenges.

## Rapid Prototyping Implementation Challenges

During the course of our research, we examined a number of GAO studies conducted over the past 12 years in which the GAO reported on the seemingly perpetual problem of technology/ prototype transition in the DoD.[1] We also interviewed Air Force representatives with

[1] GAO studies we examined for this project included *Defense Technology Development: Management Processes Can Be Strengthened for New Technology Transition Programs*, Washington, D.C., GAO-05-480, June 17, 2005; *Best Practices: Stronger Practices Needed to Improve DoD Technology Transition Processes*, Washington, D.C., GAO-06-883, September 14, 2006; *Best Practices: Stronger Practices Needed to Improve DoD Technology Transition Processes*, Washington, D.C., GAO-06-883, September 14, 2006); *Defense Technology Development: Technology Transition Programs Support Military Users, but Opportunities Exist to Improve*

experience in transitioning S&T projects into acquisition programs, as well as defense industry representatives with considerable experience in both developing and prototyping space systems.[2]

Based on this research, we found that many DoD S&T activities successfully delivered a capability to a military user, transitioned a technology to an acquisition POR, or provided a capability to a prime contractor who commercialized it in some manner for introduction into a commercial market.

However, we also identified a number of common challenges these projects have encountered when attempting to transition a working prototype or experimental capability into operational service or into an acquisition POR.[3] These include

- insufficient technical maturity or systems engineering/manufacturing robustness to support sustained military operations, that is, not producing a fieldable prototype
- lack of warfighter commitment to accept unproven capabilities into operational use or sponsor and fund their continued development
- lack of a product support infrastructure and elements, such as spare and repair parts, maintenance technical data, and peculiar ground support equipment needed to sustain the capability
- lack of validated operational tactics, techniques, and procedures to enable and sustain effective military operations
- lack of resources and materials to support initial and recurring operator and maintainer training and qualification programs

---

*Measurement of Outcomes*, Washington, D.C., GAO-13-286, March 7, 2013; *DoD Rapid Innovation Program: Some Technologies Have Transitioned to Military Users, but Steps Can Be Taken to Improve Program Metrics and Outcomes*, Washington, D.C., GAO-15-421, May 7, 2015; *Defense Advanced Research Projects Agency: Key Factors Drive Transition of Technologies, but Better Training and Data Disseminations Can Increase Success*, Washington, D.C., GAO-16-5, November 18, 2015; *Weapon Systems: Prototyping Has Benefited Acquisition Programs, but More Can Be Done to Support Innovation Initiatives*, Washington, D.C., GAO-17-309, June 27, 2017.

[2] We held discussions with staff from OTI, SMC, Assistant Secretary of the Air Force for Acquisition, Technology, and Logistics, Science and Technology, and Engineering Directorate (SAF/AQR), Boeing, and Orbital ATK.

[3] GAO, 2013; GAO, 2015a; GAO, 2015b; GAO, 2017; SAF/AQR staff, 2017; Orbital ATK staff, 2017; and Boeing staff, 2017.

- mismatches between technology development and demonstration time lines and user need-by dates
- changing user technology requirements, often due to evolving combat/contingency operations that alter technology demands over time
- competition for scarce resources and delays in securing funding to transition a prototype capability into an acquisition POR due to the lengthy DoD budget preparation process
- evaluating prototypes in controlled test environments rather than actual operational environments and failing to involve operators/end users in testing
- reluctance within the acquisition community to take prudent risks associated with early fielding of prototype capabilities.

### Air Force Challenges

In the course of our research, we discovered several gaps between Air Force policy and practice regarding technology/prototype transition. For example, technology transition plans are rarely prepared and only in situations where approved advanced technology demonstrations enjoy strong user support. Additionally, technology transition plans are nonbinding and only represent good faith efforts to be conducted by stakeholders involved in an S&T activity, which may have contributed to the fact none of the four Flagship Capability Concepts[4] created by the Air Force successfully transitioned from the S&T community into a POR or resulted in a fielded capability. Echoing our earlier discussion regarding general impediments to technology/prototype transition within DoD, we independently confirmed the principal reasons why many Air Force S&T initiatives seem to struggle to make the transition into acquisition PORs and into operational use:

- user skepticism and lack of commitment to transition and accept unproven technologies

---

[4] SAF/AQR staff, 2017; Per Air Force Instruction 61-101, *Management of Science and Technology*, Air Force Flagship Capability Concepts involve a demonstration of the maturity and potential of advanced technologies for enhanced military operational capability or cost effectiveness.

- lack of flexible funding streams to move innovative technologies quickly from laboratory testing to field use
- pressure on the Air Force's budget, where anything new must compete for funding against legacy programs and activities with a proven history of success.

With respect to the issue of funding to support timely technology/prototype transition, prototype development and testing can be readily funded within numerous existing Air Force Research, Development, Test, and Evaluation (RDT&E) program elements and budget activity codes. The major hurdle seems to be securing the O&S funding needed to deploy, operate, and sustain a fieldable prototype as soon as development and testing efforts are completed. Moreover, the time lags associated with the Air Force/DoD programming and budgeting process are particularly problematic.

Our research uncovered the fact that *rapid prototyping* is not generally accompanied by *rapid test and evaluation* and *rapid fielding and sustainment*, creating slack time in the prototype development and evaluation process and delaying the transition into operational service. The Air Force test and evaluation community is neither well structured nor funded to support rapid prototype test and evaluation. Nor is the Air Force S&T community well suited to provide the robust sustainment of a prototype system or component beyond that which is necessary to operate the capability in laboratory or controlled test environments. Our discussions with Air Force representatives also revealed that product support and sustainment requirements are rarely if ever considered during the development of prototype systems or components.[5]

## Overcoming Challenges to Implementing Rapid Prototyping

The technology/prototype transition challenges described in the previous section are characteristics of a chasm that often develops between a military service's S&T community and its acquisition and operational warfighting communities. For decades, this chasm has been termed

---

[5] SAF/AQR staff, 2017.

the "technology valley of death." The military's primary valley of death challenge is to *operationalize* a new technology or prototype capability such that an end user can quickly introduce and effectively employ it on the battlefield and provide an enduring, sustainable warfighting capability. As noted earlier in this chapter, history suggests first-of-a-kind prototype systems, subsystems, or components rarely transition from an experimentation activity directly into operational service. For reasons we identified in the previous section—lack of technical maturity/robustness, logistics infrastructure, funding, various generic impediments to transition, and so on—AFSPC may face similar challenges when attempting to transition a rapid prototype directly into operational service.

Based on our literature review and interviews with government and defense industry subject-matter experts, we developed two overarching recommendations AFSPC should consider to overcome the challenges of operationalizing a rapid prototype:

- institutionalize a disciplined, repeatable, and measurable prototype development, evaluation, and transition process
- associate and integrate independently conducted rapid prototyping activities with an acquisition POR as early as possible.

These overarching recommendations and specific supporting recommendations are summarized in Table 4.1 and indicate which apply to AFSPC, SMC, or both.

### Overarching Recommendation 1: Institutionalize a Prototype Development, Evaluation, and Transition Process

Prototype systems are typically not designed to provide a long mission life, are generally not subject to rigorous developmental and qualification testing, and often do not have a mature product support structure in place. Rapid prototyping will require prudent risk-taking in each of these characteristics to speed acquisition decisionmaking and rapidly integrate new capabilities into evolving space architectures. To manage the risks associated with rapid prototyping, AFSPC should consider institutionalizing a disciplined, repeatable, and measurable prototyping development, evaluation, and transition process. Our high-level concept for such a process includes a series of streamlined events starting

**Table 4.1**
**Recommendations for Rapid Prototyping**

| Recommendation | AFSPC | SMC |
|---|---|---|
| *Overarching Recommendation 1: Institutionalize a Prototype Development, Evaluation, and Transition Process* | | |
| Adopt and employ technology/prototype transition best practices | X | X |
| Codify governance, process, and agreements for prototype transition | X | X |
| Track transition readiness using USSOCOM's Transition Confidence Level (TCL)[a] | X | X |
| *Overarching Recommendation 2: Associate and Integrate Rapid Prototyping with Acquisition PORs* | | |
| Integrate rapid prototyping activities with operational plans | X | |
| Integrate rapid prototyping activities with acquisition POR content, ensuring JCIDS artifacts, Program Objective Memorandum (POM) budget requirements, and acquisition program strategies and plans are addressed | X | X |
| Leverage recent NDAA provisions related to rapid prototyping | X | X |

[a] Anthony Davis and Tom Ballenger, "Bridging the Valley of Death," *Defense AT&L Magazine*, January–February 2017, pp. 13–17.

at the point where a rapid prototyping project comes into existence, progresses through the experimentation stage, and reaches the point where a determination is made regarding disposition of the prototype system or capability. We depict these events in Figure 4.1.

In our concept, the first step in the prototype development and evaluation process (outlined in red) is the identification of a prototype developer. According to our discussions with AFSPC, the prototype developer could be any one of several organizations with equities in space-related technologies and experimentation, such as the AFRL, the AFRCO, the DoD Operationally Responsive Space office, a SMC program management office, or a commercial industry provider. Once the prototype system or capability is designed and built, it would then undergo an operational evaluation to assess its technical performance, operational effectiveness, and suitability to enter operational service or transfer into an acquisition POR for further maturation and subsequent fielding. Upon completing the operational evaluation, the prototype would be subject to a formal decision process to determine its

**Figure 4.1**
**Prototype Development, Evaluation, and Disposition Concept (Prototyping Occurring Outside the POR)**

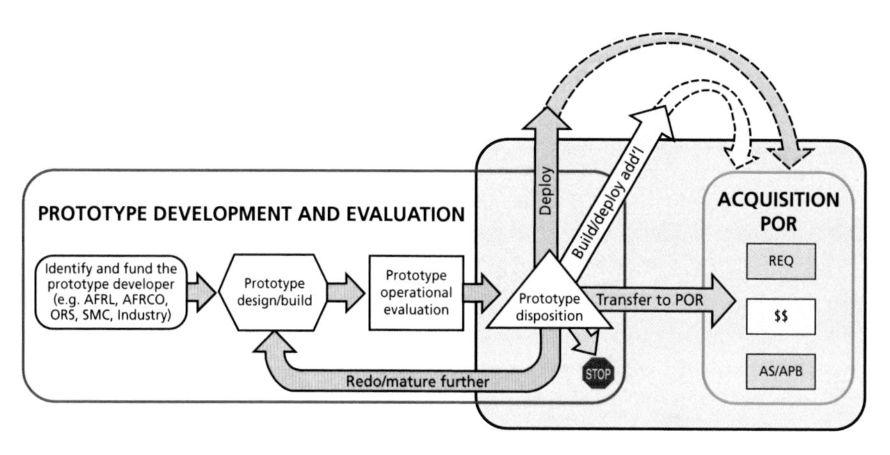


NOTE: AS = Acquisition Strategy; APB = Acquisition Program Baseline.

disposition. We envision five potential outcomes at the prototype disposition decision event:

- deploy the prototype system/capability as-is into operational service
- build additional prototype systems/capabilities and deploy them into operational service
- transition the prototype system/capability into an acquisition POR without immediately entering operational service
- stop further development and evaluation of the prototype system/capability altogether, implying it failed to achieve its intended level of performance or was deemed ineffective or unsuitable for military application
- return the prototype system/capability to its developer for additional development and potential reevaluation, implying the prototype has military utility but needs additional maturation before entering operational service or transitioning to an acquisition POR.

For prototype systems/capabilities that enter operational service directly from the prototype development and evaluation process, we recommend they subsequently transition into or be assimilated by an acquisition POR at an appropriate point in time (as depicted by the dashed arrows in Figure 4.1).

### *Adopt and Employ Technology/Prototype Transition Best Practices*

The literature we reviewed provided general recommendations for bridging the valley of death and maximizing the chances for success when transitioning a new technology or prototype capability from an experimentation environment into immediate operational use or into an acquisition POR. For example, in a 2013 report on defense technology development, the GAO stated:

> Programs must have processes in place that support the selection of projects that have realistic schedule and cost expectations, as well as sufficiently mature technologies. In addition, early project endorsement from intended users and other key stakeholders, including senior leadership and S&T developers, is critical. Once selected, projects require effective management to ensure technology risks are minimized, costs and schedules are maintained, and transition commitments are confirmed. To achieve this, stakeholders must engage in regular communication during project execution. Formal agreements and other project assessment measures also are important to ensure projects stay on track and stakeholders sustain their commitments to transition.[6]

Our literature review and interview findings further supported many of these hallmarks of successful technology/prototype transition, specifically noting the use of mature technologies, strong user commitment, and proactive program management as key factors that can contribute to a successful technology/prototype transition from an S&T activity into an acquisition program or operational use.[7] Table 4.2 summarizes these best practices regarding technology/prototype transition that are applicable to AFSPC's rapid prototyping activity. AFSPC

---

[6] GAO, 2013.

[7] GAO, 2015b; OTI staff, discussion with authors on prototyping, Arlington, Va., March 3, 2017a; Davis and Ballenger, 2017, pp. 13–17; U.S. Navy, *Future Naval Capabilities Guidebook 2017*, Arlington, Va., 2017; SMC/AD staff, 2017; SAF/AQR staff, 2017.

**Table 4.2**
**Technology/Prototype Transition Process Best Practices**

| Factor | Characteristics Supporting Successful Transition to Operational Use or an Acquisition POR |
|---|---|
| Demand signal | • There is agreement between the technology/prototype developer and a military service or warfighter representative that a military capability gap and operational requirement exists<br>• Where applicable, indications from industry partners suggest a commercial demand for the technology or an interest in commercializing it[a] |
| S&T stakeholder interest | • The technology is reflected in S&T stakeholder research-and-development activities over time<br>• The technology is linked to or evolved from prior research or related S&T projects |
| Prototype project goals | • Operational capability needs, technical performance specifications, funding requirements, and development/delivery schedules are documented and agreed to by all stakeholders<br>• Technical goals are testable and measurable<br>• Pass/fail or other success criteria are identified and agreed to by all stakeholders |
| Technical maturity | • Prototypes have a minimum Technology Readiness Level (TRL) of 6 or greater[b]<br>• Government subject-matter experts have reviewed and agree with the technical approaches and TRL information provided by potential prototype developers |
| User commitment | • A specific transition partner (operational end user or acquisition program manager [PM]) and prototype development, evaluation, operation, and sustainment funding requirements and potential sources are identified prior to prototype selection and demonstration<br>• Potential/identified end users are tightly integrated with prototype development, demonstration, and evaluation activities<br>• User/acquisition PM interest and commitment is retained as the prototyping activity evolves (else consider stopping the work if interest wanes) |
| Program management | • Specific roles and responsibilities are identified for prototype project participants<br>• Frequent project reviews are conducted to track progress and correct project course as necessary<br>• Prototype transition plans, events, and success criteria are codified in a transition agreement<br>• If immediate operational deployment and/or follow-on production is envisioned, the prototype vendor is able and committed to sustain the prototype(s) and/or ramp up production of additional prototype systems as necessary to achieve operational goals |

SOURCES: GAO, 2013; GAO, 2015b; GAO, 2017; U.S. Navy, Office of Naval Research, 2017; Davis and Ballenger, 2017; OTI staff, 2017a; SMC/AD staff, 2017; SAF/AQR staff, 2017.

[a] There are common interests between DoD and commercial sector in certain space capabilities and technologies.

[b] The DoD TRL scale ranges from TRL 1 (basic principles observed and reported) to TRL 9 (actual system proven through successful mission operations). TRL 6 is defined as a system/subsystem model or prototype that has been demonstrated in a relevant environment. TRL 7 is defined as a system prototype that has been demonstrated in an operational environment. TRL 8 is a defined as an actual system that has completed development and been qualified through formal test and evaluation.

NOTE: This list heavily relies on GAO reports on technology transition. Many of these practices are consistent with our interview and additional literature findings.

should consider implementing these best practices in its rapid prototype development, evaluation, and transition process and policy documents.

#### *Codify Governance, Process, and Agreements for Prototype Transition*

Given the Air Force's limited success with technology transition plans, we looked across DoD to identify potential processes and management techniques that could support the rapid fielding of prototype systems and capabilities into operational service and into acquisition programs. Our research led us to the U.S. Navy's FNC program.

The Navy instituted the FNC program in 2002 to develop and transition leading-edge technologies into ship and aircraft platforms within two to five years from project initiation. The FNC process provides a way for the Navy to insert relatively mature technologies into acquisition PORs that can field new capabilities on an accelerated time line compared to traditional acquisition processes. The FNC process is governed by a set of business rules that require frequent interaction and coordination between Navy and Marine Corps operating forces, the Department of the Navy requirements and resource management communities, its S&T community, and its acquisition community.[8]

We found the FNC Technology Transition Agreement (TTA) process and template (see Appendix F in the online appendixes) to be particularly good examples of a methodology that could assist and enable the timely transition of AFSPC rapid prototypes into evolving space architectures and orbital platforms or ground control stations on an accelerated time line. For example, space architectures and platforms are in many respects analogous to Navy ship-borne systems that could incorporate rapidly prototyped components that utilize common platform interface standards. In our view, the Navy TTA process places greater emphasis on advance planning of a technology/prototype transition than the Air Force's Technology Transition Plan process.[9]

---

[8] For additional information about the Navy FNC program, consult the FNC Guidebook (Office of Naval Research, 2017).

[9] This is supported by the GAO's consistent and positive references to the Navy's approach to technology transition in the reports we examined, some of which specifically highlight the Navy FNC program as a DoD best practice.

All FNC projects require a signed TTA from start to finish. These agreements require commitment from a resource sponsor, the FNC project manager, and the acquisition program manager who will receive the technology project. They are reviewed annually to ensure each stakeholder remains committed to the effort and its intended outcome. In general, FNC TTAs are designed to address many of the key characteristics of successful technology/prototype transition we outlined in Table 4.2.

The Navy FNC TTA template is detailed and prescriptive regarding technology transition requirements and plans, acquisition POR integration requirements, and program funding to support both the FNC project and the acquisition program receiving the technology project. The Navy's TTA guidance suggests such agreements should address the following attributes:[10]

- the technology development pathway described in detail
- expected outcomes of the project stated as measurable and achievable exit criteria
- identification of the customer for the technology
- a funding strategy that establishes the resources to be provided by source, appropriation, amount, and timing for technology development and follow-on acquisition and fielding activities
- a detailed project schedule with identifiable milestones, including the transition or technology handoff schedule
- the acquisition strategy and integration plan for adopting and integrating the emerging technology
- issues and risks, such as cost, schedule, technical, manufacturability, and sustainment
- customer and program manager agreement (signatures) regarding funding, schedule, and FNC project deliverables.

Another positive attribute we discovered with the Navy's FNC program is its annual Transition Assessment Review process. Each August the Office of Naval Research gathers FNC project managers,

---

[10] U.S. Navy, Office of Naval Research, "Elements of a Technology Transition Agreement (TTA)," undated.

resource sponsors, acquisition program managers, and other stakeholders to review the status and way ahead for each FNC project. The purpose of these reviews is to ensure that

- technology development activities are on track and within acceptable risks
- the transition strategy is viable
- transition planning is on track according to the TTA
- transition funding is aligned to support integration into the POR receiving the technology project/prototype and/or deployment to operational forces.

During these reviews, Office of Naval Research project managers and resource sponsors discuss each TTA and, if issues are identified, work together to resolve or mitigate concerns and revise the TTA accordingly. Resource sponsors then must provide a recommendation to the Navy Technology Oversight Group, which oversees the FNC program regarding whether to continue or cancel each FNC project. AFSPC should consider the use of signed TTAs for its rapid prototyping projects and implement a prototyping governance process/structure similar to the Navy Transition Assessment Review process.

#### *Track Transition Readiness Using a Metric*

USSOCOM is widely acknowledged for its ability to rapidly develop, acquire, and field new and innovative capabilities to its warfighters—especially those that leverage existing technologies or add improved capabilities to existing platforms and systems. We investigated its approaches and methods for transitioning emerging technologies from the command's S&T portfolio into its acquisition PORs and ultimately into its special operators' hands. In 2014, the USSOCOM Component Acquisition Executive, frustrated by chronic shortcomings in his organization's ability to transition new capabilities quickly and successfully from his S&T portfolio to his acquisition programs (an intraorganizational transition), tasked his S&T Director to develop a methodology to mitigate USSOCOM's "valley of death" problem and increase the probability of successful technology transition into acquisition and subsequent operational service. The S&T Director searched for an appropriate tool that would be relatively easy to implement, minimize cultural resistance, balance opportunities for innovation with operational necessities, and fit USSOCOM's funding realities. His solution: the USSOCOM TCL scale.

Similar to the TRL scale, the SOCOM TCL scale is a 1–9, accomplishment-based road map that plots the transition path for an emerging technology or prototype capability from complete uncertainty (TCL 1) to completed transition (TCL 9), as shown in Table 4.3.[11]

**Table 4.3**
**USSOCOM Transition Confidence Level**

| TCL | Characteristics |
|---|---|
| 1 | • Acknowledged capability/technology gap<br>• Technological discovery underway<br>• S&T working group interest expressed |
| 2 | • S&T project initiated<br>• TRL goals established |
| 3 | • S&T project technical goals established<br>• Target acquisition program(s) identified<br>• Potential transition stakeholders identified |
| 4 | • Relevant acquisition program(s) named<br>• Key transition stakeholders named<br>• Target acquisition PM(s) briefed and receiving program updates |
| 5 | • Active communication with named acquisition PM(s)<br>• Transition interest expressed by acquisition PM(s) |
| 6 | • Transition technical goals approved by S&T and acquisition PM(s)<br>• Estimated transition schedule completed<br>• S&T project included in acquisition program plan |
| 7 | • S&T project integration strategy defined in acquisition plans<br>• Transition cost estimate completed<br>• Potential funding sources identified |
| 8 | • Signed transition agreement between S&T and acquisition PM(s)<br>• Transition funding committed |
| 9 | • Transition to acquisition program and funding completed<br>• Transition after action report and storyboard documented<br>• Transition success reported to SOCOM Acquisition Executive |

SOURCE: Davis and Ballenger, 2017.

[11] Davis and Ballenger, 2017, p. 15.

Within USSOCOM's acquisition organization, the TCL scale is tailored to fit organizational objectives and the dynamics between its S&T community and acquisition community. The S&T community can advance a technology project to TCL 4 with USSOCOM RDT&E funding and with transition stakeholders merely identified. Advancement to TCL 5 and beyond, however, requires explicit coordination and cooperation with the acquisition PM who will receive the technology project. At TCL 7 and 8, senior leaders across the command are brought into the discussion to secure high-level interest and commitment to the handover from S&T to acquisition. The USSOCOM TCL scale provides a consistent, objective methodology for tracking technology/prototype development progress and readiness for transition, and AFSPC should consider incorporating this tool in its rapid prototyping process.

### Overarching Recommendation 2: Associate and Integrate Rapid Prototyping with Acquisition Programs of Record

AFSPC's plan for rapid prototyping activities will be, for the most part, conducted external to an acquisition POR. Throughout the course of our research, though, every government and industry expert we consulted regarding the subject of technology/prototype transition discussed the need to associate a prototype system or capability with an acquisition POR as early as possible to provide a "home" for the prototype and advocate for the funding needed to transition, field, and sustain the capability. To understand how AFSPC might structure a rapid prototyping activity around an acquisition POR, we considered how these activities might relate to the traditional and accelerated acquisition models described in DoDI 5000.02. Figure 4.2 graphically depicts these potential relationships as they relate to three generic acquisition paths: starting within a traditional acquisition POR, starting outside a POR, or as evolved from an urgent capability acquisition (Enclosure 13 of the DoDI 5000.02).

Common to each path is the expectation that some form of operational capability requirement will trigger an activity that could involve a rapid prototyping event. These triggers could take the form of a traditional JCIDS artifact, such as an ICD, a Capability Development Document (CDD), or a Capability Production Document (CPD). Addi-

**Figure 4.2**
**AFSPC/SMC Prototype Development and Transition Paths**

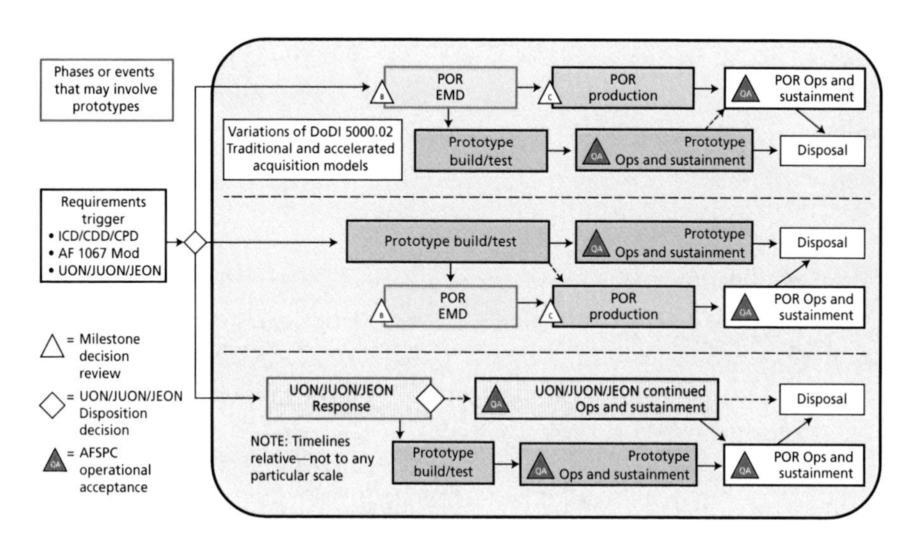


NOTE: Additional variations of these paths are possible.

tionally, an Air Force Form 1067 "Modification Proposal" could be used to initiate a modernization program for an existing fielded space system or capability. Finally, an approved major command urgent operational need (UON) or combatant command joint urgent operational need (JOUN)/JEON could initiate an acquisition activity that might involve prototyping at some point.

Referring to Figure 4.2, the top path represents a traditional acquisition POR—in this case, starting at Milestone B, the entry to the Engineering and Manufacturing Development (EMD) phase of a POR—where at some point during the POR EMD phase a particularly interesting or promising technology or capability "spins off" from the POR and a system or subsystem prototype is concurrently developed and introduced into operational service. In this example, the companion POR continues along the traditional acquisition path into the production and O&S phases and serves as a home for the operational prototype, potentially reabsorbing the prototype into the POR technical baseline at some future point in time.

The middle path represents a situation where a requirement trigger spawns a rapid prototype development and evaluation activity external to an existing acquisition POR. We envision two potential subpaths for the prototype in this scenario. In the upper subpath, the prototype system or capability is deemed sufficiently effective and suitable to support immediate fielding, and it goes into operational service immediately. The lower subpath represents a technology insertion, where the prototype system or capability merges into an acquisition POR initiated at some point after prototype development has begun. In this latter case, the POR would assimilate the prototype capability into its technical baseline during the EMD or production phases. In our view, the rapid prototype could follow both subpaths concurrently.

The bottom path represents a typical UON/JUON/JEON scenario, where some form of materiel capability is quickly fielded to deal with an urgent battlefield need. As previously noted, such materiel responses generally take the form of a fairly mature "off-the-shelf" solution with minimal development, integration, and testing. Normally, the quantity and capability of UON/JUON/JEON solutions are limited to the absolute minimum needed to mitigate the combat deficiency. However, such scenarios could spawn a follow-on activity to further mature the UON/JUON/JEON solution into a more robust or operationally capable prototype that can proceed directly into operational service. Such prototypes could remain in service on their own accord or, as shown in Figure 4.2, be merged into a follow-on acquisition POR for enduring sustainment of the capability.

As discussed in the "Rapid Prototyping Implementation Challenges" section at the beginning of this chapter, mismatches between technology development time lines and user need-by dates, changing user priorities/needs, and delays in securing funding to transition a prototype capability have historically contributed to unsuccessful prototyping activities. For rapid prototyping to succeed, planning for immediate operational deployment or transition into an acquisition POR ideally should be accomplished concurrently with prototype development and evaluation activities. Conducting these actions *after* the prototype is evaluated and deemed fit for operational service or transition would otherwise introduce unnecessary and potentially unacceptable delays in fielding the capability.[12]

Our recommendation in this regard is to assume success and plan accordingly. Predetermine the intended outcome(s) for the prototype, and document these expectations in a transition agreement. Assume the prototype will be developed on time, and provide the intended operational performance or functionality needed to address its underlying capability requirement. Assume the prototype will successfully complete its operational evaluation and be deemed effective and suitable for operational deployment, or transition to an awaiting acquisition POR. The Navy's FNC project TTA and USSOCOM's TCL process/scale may serve as useful tools and techniques to begin the prototype integration process.

#### *Integrate Rapid Prototyping Activities with Operational Plans*

Regarding planning for the operational deployment of a prototype, AFSPC fortunately has an established process in place to field new equipment and capabilities: its Operational Acceptance Process, as codified in AFSPC Instruction (AFSPCI) 10-605.[13] This instruction provides a comprehensive tool to ensure a prototype system or capability is postured to enter operational service, particularly if the Mission Assignment Process for experimental activities detailed in Attachment 6 of the instruction is followed. We consider this instruction ideally suited to accomplish operational integration and see no reason to modify or append this policy.

#### *Integrate Rapid Prototyping Activities with Acquisition POR Content, Ensuring JCIDS Artifacts, Program Objective Memorandum/Budget Requirements, and Acquisition Program Strategies and Plans Are Addressed*

Regarding transitioning a prototype system or capability to an acquisition POR, our recommendations are as follows:

---

[12] Such is the case with the ESBMC2 program we discuss in Chapter 5 of this report.

[13] AFSPCI 10-605, *Operational Acceptance Process*, June 2016.

- Predetermine where (the milestone or phase) and when the prototype will merge with the POR, and develop a planning horizon (activities, events, time lines, decision points, etc.) backward from that point.
- Ensure this planning horizon accommodates the major aspects that constitute a POR:
  - development/modification and approval of the JCIDS artifact (CDD, CPD) appropriate for the planned POR entry point
  - development and approval of Air Force POM funding initiatives and/or President's Budget content to
    - integrate the prototype system or capability into POR physical and functional baselines
    - modify, mature, or modernize the prototype to meet POR objectives
    - produce and field additional prototype systems or capabilities
    - operate and maintain the prototype system(s) or capability(ies) over its (their) intended life span
    - develop/modify and implement operator and maintainer training curricula, devices, and other training-related requirements
    - implement any required facility modification or construction requirements related to the prototype system or capability.
  - development and approval of changes to acquisition program strategies, plans, and documents necessary to integrate or accommodate the prototype system or capability, such as the acquisition strategy, acquisition program baseline, life-cycle sustainment plan, systems engineering plan, and test and evaluation master plan.
- Assign representatives from the prototype development and evaluation team to liaise with counterparts in the acquisition program management office receiving the prototype to coordinate the transition.
- Assign representatives from the program office receiving the prototype to participate in prototype development and evaluation

activities and conduct early acquisition program planning and coordinate the transition.

### *Leverage Recent National Defense Authorization Act Provisions Related to Rapid Prototyping*

In the FY 2016 and FY 2017 NDAAs, Congress has used its legislative powers to enact several statutory provisions that can enable AFSPC and SMC rapid prototyping in support of SEV implementation. These provisions provide new authorities to DoD and military services that can be used to streamline prototype contracting, funding, development, and fielding processes and accelerate the fielding of innovative technologies and prototype capabilities.

We detail these recent statutory provisions in Appendix G, but the most relevant ones for AFSPC and SMC appear below:[14]

- FY 2016 NDAA Section 804 establishes statutory direction for "middle tier" acquisition pathways that would be completed between two and five years from inception (see notes under 10 USC § 2302 for the latest version, as amended). The act specifically requires the Secretary of Defense to establish guidance for two acquisition pathways: rapid prototyping and rapid fielding. With rapid prototyping, Congress's intent is to provide a means to rapidly develop fieldable prototypes that demonstrate innovative technologies and new capabilities in response to emerging military needs. With rapid fielding, Congress's intent is to provide an ability to rapidly produce and field new or upgraded systems with minimal development requirements. The most beneficial aspect of this legislation is that it specifically exempts (except to the extent specifically provided in the secretary's guidance) rapid prototyping and fielding programs from having to comply with JCIDS Manual and DoDD 5000.01 requirements and instead calls for new processes to be put in place that enable a rapid proto-

[14] See 10 USC [United States Code] § 2302 notes; 10 USC § 2447a–e; 10 USC § 2371b.

typing or fielding requirement to be approved within six months from the time such processes are initiated.

- The FY 2017 NDAA Section 806 further strengthens the rapid prototyping and rapid fielding provisions enacted the previous year by adding new sections 2447a–e to 10 USC. These new statues include requiring that weapon system component or technology prototype projects be completed within two years of initiation and that they have a plan to transition the technology from the prototype into a fielded system, a POR, or into operational use upon successful achievement of project goals; authorizing prototype projects to be funded via traditional FAR-based contracts, cooperative agreements, or Other Transaction Authority (OTA) (10 USC § 2371); authorizing the service acquisition executive of each military department to award a follow-on production contract for prototype technologies without the use of competitive procedures under certain conditions; and authorizing the secretary of each military department to transfer unobligated procurement funding as needed to fund low-rate initial production of systems or components for rapid fielding projects (within certain limitations). See 10 USC § 2447a–e for these statutes and any amendments.

## Further Considerations for Air Force Space Command and Space and Missile Systems Center

Rapid prototyping has the potential to field capabilities in a responsive manner, but there are several key challenges in rapidly transitioning prototypes to operations or acquisition. These challenges, however, could be managed with the appropriate level of advance planning and coordination, and we discussed several best practices that AFSPC might consider adopting to improve the likelihood of successful rapid transition. That said, there are implications and trade-offs that AFSPC should consider when determining the scope of rapid prototyping (e.g., the number of prototypes to fund and field or for which capabilities prototypes should be built).

For one, many activities involved in advance planning and coordination assume that every prototype would be successful. These activities (e.g., developing JCIDS artifacts, concept of operations, training material, etc.) can be time-consuming and have resource implications. Given that the intent of prototyping is to experiment to determine or demonstrate operational utility, not all prototypes will succeed or warrant transition to operations or acquisition, and the advance planning work done for prototypes whose eventual outcome is disposal would go unused.

Additionally, rapid prototyping may provide a less-than-optimal solution to a capability shortfall or gap. For instance, in the context of AFSPC's rapid prototyping, very novel or new technology may be excluded for consideration to reduce risk and to accelerate the fielding of a capability. Further, prototypes may pose greater operational risk when they transition to operations because they are not designed to meet life-cycle supportability requirements (such as reliability and cybersecurity requirements). Should AFSPC and SMC decide to pursue increased reliability or cybersecurity measures for prototypes, such efforts may come at the expense of speed. That said, providing an optimal solution too late to address emerging threats also poses an operational risk in the context of the SEV.

CHAPTER FIVE

# Case Study: Enterprise Space Battle Management Command and Control Program

In Chapters 2 through 4, we examined effective ways to employ MOSA, agile methods, and rapid prototyping in a DoD setting. In this chapter, we provide concrete examples of how some of those recommendations might be implemented in a specific program: the ESBMC2 program. The ESBMC2 program has characteristics and goals that align well with the principles of MOSA, agile, and rapid prototyping as we will discuss later in the chapter. Because space BMC2 capabilities are critical components to realizing the SEV, ESBMC2 is an important program to AFSPC and the broader national security space enterprise. This program is in an early phase of the program life cycle, and thus the opportunity for AFSPC and SMC to apply the recommendations discussed in Chapters 2 through 4 is timely.

This chapter begins with an overview of the ESBMC2 program to provide the context within which MOSA, agile, and rapid prototyping would be employed. We then highlight an initial set of implementation actions for AFSPC and SMC's consideration based on the research findings presented in Chapters 2 through 4. These suggested actions are not based on a comprehensive evaluation of the program because we have limited information about the program given its sensitive nature and early stage in the program life cycle.

## Enterprise Space Battle Management Command and Control Context

The Joint Functional Component Command for Space (JFCC Space),[1] a component of U.S. Strategic Command (USSTRATCOM), uses multiple operating systems that provide space situational awareness (SSA) capabilities and some support to space battle management C2. To replace the legacy space operating systems and provide additional capabilities to meet broader ESBMC2 needs, the Air Force began development of JMS in 2010 in successive increments. JMS Increment 1 delivered an initial service-oriented architecture and user tools to access and analyze data from legacy systems. Increment 2, currently in development and testing, will replace the legacy space catalog system to provide data required for C2 of the national security space systems.[2] JMS Increment 3 was renamed in FY 2017 as ESBMC2, and this program will deliver an integrated space operating system to provide national security space enterprise with BMC2 services, including threat indications and warnings, spectrum common operating picture, modeling and simulation for course-of-action development, and other planning and C2 capabilities.

### Scope of Enterprise Space Battle Management Command and Control

The ESBMC2 system will support two operation centers performing elements of the space BMC2 mission set—the Joint Space Operations Center (JSpOC) and the National Space Defense Center (NSDC).[3] The JSpOC is the operations center for the commander of JFCC Space. Its purpose is to provide a focal point for the operational C2 and employ-

---

[1] Since the completion of the report, USSTRATCOM restructured its components and created and inactivated JFCC. In its place, USSTRATCOM created Joint Force Space Component Command (AFSPC Public Affairs, 2017).

[2] Specifically, JMS Increment 2 is planned to provide the near real-time, high-accuracy position of all objects tracked by SSA sensors. It will also provide information on the detection and situational awareness of space events, such as launches, maneuvers, close-proximity operations, breakups, separations, reentries, conjunctions, and deorbits.

[3] Joint Force Space Component Command transitioned the JSpOC to a Combined Space Operations Center in July 2018 (Joint Force Space Component Command Public Affairs, "CSPOC Established at Vandenberg AFB," July 24, 2018).

ment of worldwide joint space forces and enable integrated space power for global military operations. DoD established the NSDC in 2015 to improve operational integration of the military and Intelligence Community's efforts to defend and secure the space domain in response to the increasing concerns about threats to space capabilities.[4]

Additionally, ESBMC2 will be integrating with other operations centers and with the systems and data sources that support those centers[5] to ensure a common operating picture to synchronize course of action (COA) planning and execution decisions. It will act as a vertical integration node at the operational level of war, interpreting and acting upon strategic guidance from the USSTRATCOM Global Operations Center and National Military Command Center while coordinating tactical actions performed by lower-echelon satellite operations centers. It may also engage in supporting/supported relationships with horizontal operation centers external to the U.S. national security space, such as theater cells in the Geographic Combatant Commander's Air Operations Centers, allied space operations centers, and commercial space operations centers. As a result, systems integration for ESBMC2 will likely be highly complex.

### Preliminary Acquisition Strategy for Enterprise Space Battle Management Command and Control

AFSPC's preliminary acquisition strategy for ESBMC2 involves leveraging other concurrent space BMC2 development efforts to accelerate the fielding of space BMC2 capabilities.[6] We discuss the program characteristics of these efforts (shown in Table 5.1)—program goals,

---

[4] U.S. Department of Defense, "New Joint Interagency Combined Space Operations Center to Be Established," press release, Release No: NR-352-15, September, 11, 2015; NSDC was renamed effective April 1, 2017, from the Joint Interagency Combined Space Operations Center (John Hyten, "Statement of John E. Hyten, Commander United States Strategic Command, Before the Senate Committee on Armed Services," April 4, 2017.

[5] For example, ESBMC2 will also integrate with other programs, such as JMS Increment 2 and Enterprise Ground System, which will receive and execute the tasking delivered from the ESBMC2 system (SMC, 2017).

[6] John W. Raymond, "Fiscal Year 2018 Priorities and Posture of the National Security Space Enterprise," Department of the Air Force Presentation to the Subcommittee on Strategic Forces House Armed Services Committee, U.S. House of Representatives, May 19, 2017.

**Table 5.1**
**Current Air Force Space BMC2 Development Efforts**

| | JEON Space BMC2 Prototype | AFRCO Space BMC2 Prototype | ESBMC2 Program of Record |
|---|---|---|---|
| **Developer** | **AFRL** | **AFRCO** | **SMC** |
| Mission/goal | Rapidly deliver initial BMC2 capabilities | Develop space BMC2 prototype using open architecture and standards | Develop an enterprise space BMC2 system with life-cycle support |
| End user | NSDC | NSDC | JSpOC, NSDC, and others |
| Time line for fielding[a] | Initial capabilities by end of FY 2018; SMC will sustain afterward | Transition to ESBMC2 POR in 2021 | Estimated MS B in FY 2019; IOC is to be determined |
| Requirements[b] | Top 25 needs for the NSDC from 2016 Space Posture Review (SPR) | Focus on infrastructure, standards, and applications for BMC2 in NSDC | To be defined (likely to include JMS 2015 IS-CDD; all needs from 2016 SPR; additional requirements arising from experimentations at NSDC and integration with other operation centers)[c] |
| Architectural framework and standards | OA-DCGS/Use of Phoenix Prime middleware[d] | Extension of OMS/UCI for space | Build upon AFRCO developed architectural framework and standards; multiple architectures and standards (e.g., JMS, EGS) |
| Reporting line/ oversight | AFMC, USSTRATCOM | Board of Directors (USD [AT&L], SECAF, CSAF, SAF/AQ) | AFPEO/SP, AFSPC |

[a] Raymond, 2017.

[b] AFRL staff, discussion on the space BMC2 JEON, 2017; U.S. Air Force, "Enterprise Space Battle Management, Command and Control System (Joint Space Operations Center Mission System Increment 3)," Report to the Congress Defense Committees, April 2017.

[c] SMC Space Defense Task Force staff, 2017.

[d] AFRL staff, 2017; SMC, Space Defense Task Force staff, discussion on ESBMC2, El Segundo, Calif., June 30, 2017.

NOTE: OA-DCGS = Open Architecture Distributed Common Ground System; MS = milestone; IOC = initial operational capability; AFMC = Air Force Materiel Command; USD (AT&L) = Under Secretary of Defense for Acquisition, Technology, and Logistics; SECAF = Secretary of the Air Force; CSAF = Chief of Staff of the Air Force; SAF/AQ = Assistant Secretary of the Air Force for Acquisition, Technology, and Logistics; AFPEO/SP = Air Force Program Executive Office for Space.

end users, program time lines, scope/requirements, architectures, and oversight—as they will affect how MOSA, agile, and rapid prototyping might be implemented in ESBMC2.

First, AFRL is developing a prototype in response to a space C2 JEON to provide initial BMC2 capabilities at NSDC. These capabilities are anticipated to be fielded by the end of FY 2018. In parallel, AFRCO has started work on the framework and a scalable space BMC2 prototype with the intent of integrating the JEON capabilities into the AFRCO prototype. The JEON prototype will transition to the ESBMC2 POR to ensure it continues to operate until it is evolved with more comprehensive and enduring solutions that meet operational standards. The AFRCO prototype will also transition to the ESBMC2 program in the 2021 time frame such that SMC can build upon the prototype and integrate it into an enterprise system with the appropriate logistics and operational support systems. The ESBMC2 program is currently in developmental planning and expects Milestone B in FY 2019.[7] The plan is for SMC to serve as the transition agent and enterprise manager for these development efforts.

The JEON and the AFRCO prototypes are focused on meeting NSDC needs and will not cover the broader space enterprise BMC2 needs. The requirements for the JEON are derived from a Space Strategic Portfolio Review in 2016 on BMC2, which documented needs for the NSDC. AFRL is initially targeting the top 25 needs in the JEON. We do not have specific information on the requirements for AFRCO's prototype, but they are focused on developing the infrastructure, standards, and software applications for conducting BMC2 in NSDC. SMC will develop capabilities that are not addressed in the JEON and the AFRCO prototype in addition to any capabilities that do not transition to the ESBMC2 system.

Multiple sets of requirements apply to the ESBMC2 POR, but a formal set of requirements is still in development. An updated 2015 JMS CDD includes new requirements that apply to BMC2 capabilities (threat indication and warning, spectrum common operating picture, and modeling and simulation for COA generation). Additionally, JFCC

---

7 U.S. Air Force, April 2017.

Space continues to define operational needs through experimentation at NSDC, and the outcome of these efforts will help refine existing requirements for the POR while identifying new BMC2 requirements to be considered.[8] It is anticipated that the requirements for ESBMC2 will evolve over time as it integrates data sources, applications, and networks with other operation centers and may continue to do so as the space threat environment evolves.

The architectures that these three separate systems are using or developing are not yet common. Given the urgency of the JEON, AFRL is taking advantage of as many existing capabilities as possible to field an initial BMC2 capability at NSDC. It is using the open architecture version of the Distributed Common Ground System (DCGS) and middleware Phoenix Prime for the JEON in its initial spiral and plans to migrate to OMS/UCI framework/standards in a later spiral.[9] AFRCO's space BMC2 standards and architecture are envisioned to be an extension of OMS/UCI. AFRCO will work with the OMS/UCI industry consortium to develop open standards for space BMC2. SMC's plan is to adopt the architecture and standards that AFRCO develops. However, it may also inherit additional legacy architectures and standards as a result of integrating with other systems (e.g., JMS and EGS) that have their own unique interfaces and data formats that are separately developed and managed.

## Opportunities to Apply Modular Open Systems Architecture, Agile Acquisition, and Rapid Prototyping

SMC is seeking flexibility and scalability of capabilities (i.e., the rapid addition of new capabilities without negatively impacting existing performance) for ESBMC2.[10] Given this end goal, SMC plans to pursue a MOSA approach for ESBMC2. Additionally, the need for interoperability among multiple centers makes MOSA an attractive strategy for ESBMC2.

---

[8] U.S. Air Force, 2017.

[9] AFRL staff, 2017.

[10] SMC, 2017.

Agile may also have a role in ESBMC2. The dynamic nature of ESBMC2 requirements, as well as the need to incrementally integrate components of ESBMC2, aligns well with the characteristic attributes of agile acquisition discussed in Chapter 3. Many separate capabilities in ESBMC2 can, in principle, be developed in separate increments or be developed in a series of increments that first achieve a minimum acceptable level of operational performance and then add performance as long as the value of doing so exceeds the cost (e.g., capabilities that SMC expect to add to the JEON and AFRCO prototype lend themselves to an agile approach).

As discussed in the previous section, rapid prototyping is already being implemented for the ESBMC2 program as AFRL and AFRCO build their prototypes to be eventually transitioned to the POR. In the future, other prototype applications may be developed outside of the ESBMC2 POR that could potentially transition to ESBMC2 (e.g., as a result of the AFSPC rapid prototyping process discussed in Chapter 4).

ESBMC2 can employ these acquisition tools synergistically. A MOSA framework would enable many of the capabilities in ESBMC2 to be developed in separate increments. The separable nature of MOSA in turn allows agile acquisition to occur within individual increments, in a series of increments, or both. Rapid prototyping can also take place within individual increments without affecting the rest of the system, and agile methods can be applied to the prototype development. These tools complement each other synergistically and help achieve the end goal of responsiveness.

### Modular Open Systems Architecture Implementation in Enterprise Space Battle Management Command and Control

While AFRCO develops a space BMC2 prototype and lays the foundation for a MOSA framework, there are many complementary activities that SMC could pursue to create an enabling environment for MOSA implementation in the ESBMC2 POR.

#### *Enterprise Governance*

As highlighted earlier, SMC will face complex technical and programmatic integration challenges as it (1) manages the transition of the JEON and AFRCO prototypes to the POR and (2) synchronizes and integrates with other systems that affect the ESBMC capabilities (e.g., JMS, EGS, or other data sources and C2 systems). The Air Force is currently establishing an ESBMC2 governance structure, which should help manage different priorities among ESBMC2 stakeholders. As the Air Force refines the membership, responsibilities, and operating principles of the ESBMC2 governance body, we suggest that the enterprise governance recommendations in Chapter 2 serve as a guide. In particular, including the following elements in the governance charter could help align the three space BMC2 efforts and prepare the POR for MOSA implementation in the near term:

- define which programs will be governed by the ESBMC2 governance (e.g., to what extent will the JEON and AFRCO prototypes be governed? Is EGS included?)
- define the authorities and responsibilities of the transition agent and enterprise manager (currently SMC)
- clarify priorities, if known, for the JEON prototype, AFRCO prototype, and the ESBMC2 POR and how they align with the enterprise objectives
- identify risk acceptance authority for the enterprise (if it is not the enterprise manager) when trade-offs are required among programs being governed
- define processes for making trade-offs among programs being governed (e.g., design decisions in AFRCO prototype that could impact ESBMC2 life-cycle supportability) and the supporting data required to inform decisions
- define processes for ESBMC2 life-cycle management decisions, such as establishing program time lines and scope/requirements for each increment, if multiple increments are planned for MOSA implementation.

#### *Up-Front Program Planning*

As the Air Force transitions ESBMC2 from a JEON effort to a formal POR, it can establish an expectation that system performance requirements will evolve over time rather than developing a single set of JCIDS documents meant to scope the entire program. This involves the incorporation of thresholds and goals, as well as design growth margin, into applicable system requirements. For example, hardware selected for the ESBMC2 POR should include design growth margin to enable additional data storage or a higher data transfer rate if a newly developed software application can take advantage of these capabilities.

Further, using a model similar to the A-RCI program, the Air Force could establish program development and fielding time lines based on expected operational needs. As with A-RCI, finalizing program time lines will require significant interaction among the requirements authority, the technical/acquisition team, and the user community through formalized processes. While adherence to formalized processes will be a key element of ESBMC2 life-cycle management, establishing metrics to evaluate the relative success of the processes is also important. Rapidly changing operational needs or other considerations[11] may require changes to programmatic time lines to meet the needs of the operational customer. While approval to employ such nonstandard acquisition methods as operating an iterative program and combining developmental and operational testing is not guaranteed, establishing an incremental process and successfully implementing early iterations of the process have enabled this approach in certain cases in the past.[12]

Two other examples of considerations that must be made early for the ESBMC2 POR are what entity will be the system integrator and what interaction/interoperability does the ESBMC2 system need with other space enterprise systems.

While ESBMC2 is still in the prototype phase, too great a focus on life-cycle support issues could derail ESBMC2 prototype technical plans. That said, a degree of early consideration for these issues could still inform design decisions to the benefit of these issues in the future.

### *Standards Development and Management*

As the space-specific version of the OMS/UCI standards are developed, continued interaction with industry through consortiums is

---

[11] See A-RCI Case Study in Appendix B.

[12] As performed by the A-RCI program. See Appendix B.

important to ensure the newest technologies continue to be considered for inclusion in the standards. If not already established, long-term standards management and governance plans must be developed for the OMS/UCI standards. Additionally, pushing for broader use of the UCI standard within the space industry would benefit its market penetration and make suppliers more interested in using it on their own. During our discussions with industry, it was identified that if industry sees the government is committed to the standard, they will be more willing to commit to it. Even if the Air Force only pushed for broader use of OMS/UCI within DoD, taking this step could provide a sufficient business case for a broader base of suppliers to adopt the standard.

On a more technical note, rather than immediately replacing all early ESBMC2 JEON efforts with OMS/UCI compliant software modules, the Air Force should consider using such techniques as middleware to connect these modules to compliant OMS/UCI buses. In accordance with MOSA and agile methods discussed in this report, a focus on the quick fielding of operational capability should be the primary driver for design decisions, provided the component can interact across the standard interface. Later iterations can focus on transitioning these modules to full compliance for improved life-cycle sustainment.

### Agile Implementation in Enterprise Space Battle Management Command and Control

In Chapter 3, we discussed the initiatives that AFSPC and SMC need to pursue to institutionalize an agile acquisition and development. This set of initiatives is a long-term effort, and thus agile acquisition is not likely to be institutionalized by the time ESBMC2 reaches Milestone B. The ESBMC2 program office would need to determine whether it is ready for agile acquisition at that time, depending on the scale of agile acquisition that SMC chooses to pursue for ESBMC2. If SMC does pursue an agile approach for ESBMC2, there are some initial steps that the program office can take to prepare for agile.

First, the program office should focus on building the team with agile experience and skills to manage an agile development. In particular, the program office should plan to recruit acquisition staff involved in the AFRL JEON and AFRCO prototype. AFRL is using an agile acquisition approach to develop the JEON capabilities because of very tight time constraints and because it provides the flexibility required to allow a largely parallel approach to occur in a way that addresses operator requirements as early and often as possible.[13] The agile development for the JEON appears to be implementing many best practices for agile, such as securing Air Force senior leadership support and active user participation from NSDC operators;[14] using contractors with experience in agile software development; and establishing appropriate testing infrastructure and practices for agile, such as automated testing, a software integration laboratory to support "sprint" testing, and other common testing practices used in agile.[15] We do not know whether AFRCO will also pursue agile methods for its prototype development, but AFRCO is being directly informed of all of AFRL activities as it prepares to transition on the JEON into a prototype. Additionally, AFRCO's own core mission is to accelerate acquisition activities, and the agile methods discussed in Chapter 3 are among the broad set of methods AFRCO has applied since the Air Force set up the office.[16]

---

[13] AFRL staff, 2017.

[14] The Air Force leadership gave the JEON more requirements flexibility than a standard Air Force program office gets. That flexibility allows the JEON to avoid delays traditionally associated with adjusting requirements under JCIDS. The leadership's perspective results in part from a recent unsuccessful experience with agile acquisition that they believe was hampered by excessive bureaucratic structures and rigidities (see Appendix D). Agile developers are currently working with users in the NSDC that will receive the product of the JEON when it is ready to be fully fielded. Operational personnel from NSDC are currently present at various development sites; development personnel are working onsite at the NSDC as it stands up (AFRL staff, 2017).

[15] Contractors are also conducting regression testing during development, a common agile practice, to ensure that new code is robust to integration with other code. More formal integration and testing occur at the close of each sprint (AFRL staff, 2017).

[16] The characteristics of AFRCO's typical acquisition approach are similar to those of agile acquisition. They include short narrow chain of command, small integrated program office teams, early and prominent warfighter involvement, fielding of key capabilities soonest, not waiting for development to field 100-percent solutions, focused oversight based on trust, and tailored acquisition planning (Walden, 2016).

The ESBMC2 program office can further augment the team with staff with acquisition experience in other BMC2-like programs that used agile approaches[17] and external experts to serve as agile coaches.

Second, the program office should advocate for an agile champion and agile advocates (see Chapter 3) in AFSPC to ensure leadership support of agile acquisition approach throughout the ESBMC2 program life cycle. The Air Force leaders currently appear open to a more complete agile approach in the JEON. The exceptional nature of and high-level interest in a JEON often makes it easier to tailor the management of a program. But experience tells us that, for agile to succeed in the long term, the leadership must maintain its active support.

Third, the ESBMC2 program office can begin identifying the appropriate user representatives for ESBMC2. We discussed the criticality of and challenges associated with user participation in agile acquisition in Chapter 3. The program office can consider various options, and they require careful evaluation for feasibility. For instance, with the diverse user communities that ESBMC2 will serve, something closer to the unaffiliated product owner team of experts may be more appropriate.[18]

In the long run, experience from ESBMC2 can feed the broader goal of institutionalizing processes to drive continuous improvement of how the government implements agile. For instance, AFSPC and SMC have provided personnel to AFRL and AFRCO to support the

---

[17] For instance, with help from the Defense Innovation Unit Experimental (DIUx) and Defense Digital Service, Raytheon is using an advanced version of agile, DevOps, to overcome earlier development problems that led to unacceptable schedule delays and cost growth in the Operational Control System program to develop the ground segment for the next-generation version of GPS (Dee Ann Divis, "OCX Passes Deep Dive Review; GAO Says Program Risk Remains High," *Inside GNSS*, May/June 2017; Valerie Insinna, "US Air Force Creates Group to Recruit Cyber Nerds for Weapons Programs," *Defense News*, January 6, 2017a). DIUx is using similar methods to upgrade tanker management in the 609th Air Operations Center (AOC), which supports U.S. Central Command from El Udeid Air Base and Shaw Air Force Base. DIUx is planning to use DevOps to pursue an AOC pathfinder project to find a better approach to developing the BMC2 capabilities sought in the now-cancelled Falconer AOC 10.2 program (Valerie Insinna, "Air Force Cancels Air Operations Center 10.2 Contract, Starts New Pathfinder Effort," *Defense News*, July 13, 2017b).

[18] GAO, 2012.

JEON and the prototype. The time they spend in these efforts should give them useful experience with agile methods, and they can bring that experience to bear in the ESBMC2 program and in other programs in the Air Force. AFSPC or SMC should make appropriate efforts to capture relevant information from ESBMC2 government personnel in a form that can be shared. In anticipation of such a long-term training effort, AFSPC and SMC can build arrangements to capture lessons learned in the ESBMC2 program and others that can inform their various personnel in future programs. These arrangements will probably demand the creation of documentation that ESBMC2 users do not value themselves. AFSPC and SMC must be the advocates for such documentation when its value is greatest outside the ESBMC2 program. They cannot rely on the users that agile emphasizes to do this.

### Rapid Prototyping Implementation in Enterprise Space Battle Management Command and Control

Given the Air Force's current plan to field and operate the AFRCO BMC2 prototype as-is, we recommend that AFSPC consider applying the principal elements of the institutionalized prototype development, evaluation, and transition process we described in Chapter 4 to the ESBMC2 effort. Specifically, we recommend AFSPC

- consider implementing the technology/prototype transition process best practices (Table 4.1) when developing and refining the AFRCO prototype development activity plans and the follow-on transition into the planned SMC ESBMC2 POR
- consider developing and implementing an ESBMC2 technology transition agreement cosigned by AFRL, AFRCO, SMC, and AFSPC, using the Navy TTA template as a starting point (see Appendix F in the online appendixes)
- consider using the USSOCOM TCL approach to track the progress and readiness of the AFRCO BMC2 prototype to transition into field use and the SMC-led ESBMC2 POR.

As AFSPC and SMC have already committed to associating the AFRCO BMC2 prototype with a follow-on ESBMC2 POR to provide a long-term home for the prototype capability, we further recommend AFSPC, with SMC assistance,

- concurrently (with the AFRCO effort) develop appropriate JCIDS artifacts and insert appropriate funding wedges into the Air Force's FY 2020 POM to integrate, operate, and sustain the AFRCO prototype in FY 2021 and beyond
- concurrently develop BMC2 fielding plans, operating concepts and procedures, operator training requirements and plans, and logistics support plans as the AFRCO prototype technical and functional capability baselines emerge.

## Summary

ESBMC2 is likely to employ MOSA, agile, and rapid prototyping to a varying degree. These acquisition tools can potentially help ESBMC2 achieve the goal of rapid acquisition and enterprise synchronization, but AFSPC and SMC can expect to face some or all the challenges discussed in Chapters 2 through 4. In anticipation of those challenges, AFSPC and SMC should prepare to manage them to realize the key benefits of MOSA, agile, and rapid prototyping approaches. In this chapter, we highlighted a sample of initial steps that AFSPC and SMC might consider. These steps should be considered in the context of other development efforts that SMC may be pursuing for ESBMC2.

CHAPTER SIX

# Concluding Thoughts

As noted at the outset of this report, the SEV was developed because the United States recognized that it no longer has unimpeded access to the space domain. The SEV describes an integrated approach to building a force across all space mission areas, coupling the delivery of space mission effects to the warfighter with the ability to protect and defend space capabilities against emerging threats. The research summarized in this report recognizes that changes to the way space systems and supporting ground elements are acquired are critical to realizing the SEV. In particular, more responsive acquisition can improve the resilience of the space enterprise.

We identified MOSA, agile acquisition, and rapid prototyping as promising acquisition approaches to support SEV objectives. While these approaches are not new concepts, DoD has struggled to implement them for a variety of reasons. As AFSPC and SMC consider these approaches in several initiatives in support of the SEV, they should be aware of these key impediments and challenges and develop policies and guidance to posture themselves for success.

## Challenges

As discussed in the previous chapters, specific challenges for these acquisition approaches in a DoD setting, such as the SEV, are as follows:

- MOSA
  - aligning diverse stakeholders when MOSA is applied to a large enterprise

  - managing acquisition through incremental development and fielding
  - determining key interfaces that should be open
  - tailoring engagement with industry on data rights and standards
  - selecting an appropriate system integration model
- agile acquisition
  - tailoring agile to specific circumstances and choosing the right hybrid
  - defining requirements to get full benefit of agile methods
  - ensuring that developers and those who oversee them have appropriate training and experience
  - maintaining timely, committed user participation
- rapid prototyping
  - securing user support and funding source(s)
  - addressing operation and acquisition considerations after the fact.

To overcome these challenges and increase the likelihood of success, we presented in the previous chapters overarching or strategic recommendations for each acquisition approach with more specific recommendations suited to aspects of implementation. The recommendations focus on careful upfront planning to establish infrastructures for MOSA, agile acquisition, and rapid prototyping, which will require AFSPC and SMC to allot appropriate resources and schedule and plan accordingly. Further, these recommendations apply to many functions and organizations (e.g., requirements, user, or testing communities) beyond the program office. AFSPC and SMC should implement these recommendations in a holistic manner and tailor them to individual program context.

## Posturing for Success

Should MOSA, agile, and rapid prototyping have a role in the SEV, there are overarching policy-level actions that senior leadership in AFSPC and SMC may need to take to better posture the space enterprise for the successful implementation of these three approaches.

### Shift in Architectures

MOSA, agile, and rapid prototyping may have had limited opportunities for application in space acquisition because of certain structural issues in the space enterprise, such as low-volume space market, long spacecraft life, or stovepipe architecture (e.g., satellite-specific ground systems). These issues need to be addressed to better posture for the successful implementation of these approaches. For instance, as AFSPC and SMC develop future architectures for the SEV, they should evaluate whether and how these factors could change in SEV implementation and consider architectures that are more conducive to MOSA, agile, or rapid prototyping, such as by including shorter life satellites, commonality, or commercial standards.

### Investments in Education and Training of the Workforce

Space acquisition has had limited experiences with MOSA, agile acquisition, and rapid prototyping, and thus the workforce is likely to be largely unfamiliar and untrained in applying these concepts to space acquisition. An effective implementation of these approaches will require investments in human capital, especially for agile acquisition. Education and training would continue to build experiences with these approaches in the workforce and enable them to identify feasible opportunities and apply these acquisition approaches effectively. Beyond the acquisition workforce, education and training need to occur writ large to include other communities that support or enable acquisition, such as the user, requirements, and testing communities.

### Shift in Organizational Culture

MOSA, agile, and rapid prototyping emphasize different aspects of acquisition performance compared to the traditional acquisition approach. For one, they emphasize responsiveness over system performance—that is, the timely delivery of a less-than-optimal solution ("80-percent" solution) would be valued more than an optimal, complete solution that is delivered too late to address an emerging threat or exploit an emerging technology. In the case of MOSA, the emphasis is also on the benefits for the enterprise rather than for an individual program. In addition to education and training, the emphasis on responsiveness or enterprise view requires a shift in organizational culture to look at risk and "requirements" differently (than in the traditional acquisition setting) across the board within the acquisition management system: from requirements generation to testing to contracting and oversight, among other functions.

## Concluding Remarks

There may be multiple opportunities to implement MOSA, agile acquisition, and rapid prototyping in the SEV as AFSPC evolves its space and ground architectures. AFSPC and SMC should determine the scope and applicability of these approaches in the SEV considering the trade-offs associated with them. While responsiveness (and enterprise synchronization in the case of MOSA) would be primary reasons for considering these approaches in the SEV, there are other trade-offs that warrant further consideration as discussed in previous chapters. Posturing for success entails big muscle movements, such as changes in the space enterprise architectures, investments in human capital, and shifting the organizational culture. Should MOSA, agile, and rapid prototyping have a role in the SEV, we suggest that formal change-management tools be employed to assist with implementation, needed collaboration, and the broader organizational changes that these alternative acquisition approaches call for.[1]

---

[1] Refer to Appendix E for details.

# References


Acoustic Rapid COTS Insertion Program Office staff, discussion on A-RCI lessons on MOSA implementation, Washington, D.C., January 31, 2017.

AFRCO—*See* Air Force Rapid Capabilities Office.

AFRL staff—*See* Air Force Research Laboratory staff.

AFSPC—*See* Air Force Space Command.

AFSPCI—*See* Air Force Space Command Instruction.

Air Force Office of Transformational Innovation staff, discussion with authors on prototyping, Arlington, Va., March 3, 2017a.

———, discussion with authors on OTA, Open Architecture, Arlington, Va., May 31, 2017b.

Air Force Rapid Capabilities Office, "Common Mission Control Center (CMCC) Program Overview," May 16, 2017, Not available to the general public.

Air Force Research Laboratory staff, discussion on the space BMC2 JEON, teleconference, May 3, 2017, and June 26, 2017.

Air Force Space Command, "AFSPC-NRO Shared Strategic, Space Enterprise Vision," briefing, July 2016, Not available to the general public.

———, "Space Warfighting Construct," briefing slides, May 2017a.

———, "Space Mission Assurance Framework," September 2017b, Not available to the general public.

Air Force Space Command Instruction 10-605, *Operational Acceptance Process*, June 2016.

Air Force Space Command Public Affairs, "AFSPC Commander Announces Space Enterprise Vision," April 11, 2016. As of May 17, 2018:
http://www.afspc.af.mil/News/Article-Display/Article/730817/afspc-commander-announces-space-enterprise-vision/

———, "AFSPC Commander Becomes JFSCC, Joint Space Forces Restructure," December 3, 2017. As of May 22, 2018:
http://www.afspc.af.mil/News/Article-Display/Article/1386530/afspc-commander-becomes-jfscc-joint-space-forces-restructure/

Air Force Space Command staff, teleconference discussion on prototyping, May 17, 2017.

Anderson, Frank J., Jr., *A Plan to Accelerate the Transition to Performance-Based Services: Report of the 912(c) Study Group for Review of the Acquisition Training, Processes, and Tools for Services Contracts*, Washington, D.C.: U.S. Department of Defense, Report AF903T1, 1999.

A-RCI—*See* Acoustic Rapid COTS Insertion Program Office staff.

Baldwin, Laura, Frank Camm, and Nancy Young Moore, *Strategic Sourcing: Measuring and Managing Performance*, Santa Monica, Calif.: RAND Corporation, DB-287-AF, 2000. As of October 24, 2018:
https://www.rand.org/pubs/documented_briefings/DB287.html

Beck, Kent, Mike Beedle, Arie van Bennekum, Alistair Cockburn, Ward Cunningham, Martin Fowler, James Grenning, Jim Highsmith, Andrew Hunt, Ron Jeffries, Jon Kern, Brian Marick, Robert C. Martin, Steve Mellor, Ken Schwaber, Jeff Sutherland, and Dave Thomas, "Manifesto for Agile Software Development," 2001. As of January 12, 2017:
http://agilemanifesto.org/

Boehm, Barry, and Richard Turner, *Balancing Agility and Discipline: A Guide for the Perplexed*, Boston: Addison-Wesley, 2004.

Boeing staff, discussion on MOSA, agile, and rapid prototyping, El Segundo, Calif., June 29, 2017.

Broadus, William, "The Challenges of Being Agile in DoD," *Defense AT&L*, January–February 2013, pp. 5–9.

Camm, Frank, "Adapting Best Commercial Practices to Defense," in Stuart E. Johnson, Martin C. Libicki, and Gregory F. Treverton, eds., *New Challenges, New Tools for Defense Decisionmaking*, Santa Monica, Calif.: RAND Corporation, MR-1576-RC, 2003, pp. 211–246. As of August 23, 2018:
https://www.rand.org/pubs/monograph_reports/MR1576.html

———, "Federal Agencies Can Adapt Best Commercial Practice to Improve Their Acquisition of Services," testimony presented to the Acquisition Advisory Panel, Santa Monica, Calif.: RAND Corporation, CT-261, 2006. As of October 24, 2018:
https://www.rand.org/pubs/testimonies/CT261.html

———, *Analysis of Technology Innovator Industrial Base for Joint Counter Radio-Controlled Improvised Explosive Device Electronic Warfare 3.3*, Santa Monica, Calif.: RAND Corporation 2011, Not available to the general public.

Camm, Frank, Laura Werber, Julie Kim, Elizabeth Wilke, and Rena Rudavsky, *Charting the Course for a New Air Force Inspection System*, Santa Monica, Calif.: RAND Corporation, TR-1291-AF, 2013. As of August 23, 2018: https://www.rand.org/pubs/technical_reports/TR1291.html

Chairman of the Joint Chiefs of Staff, *Manual for the Operation of the Joint Capabilities Integration and Development System (JCIDS)*, February 12, 2015, including errata as of December 18, 2015. As of October 14, 2018: http://www.acqnotes.com/wp-content/uploads/2014/09/Manual-for-the-Operationsof-the-Joint-Capabilities-Integration-and-Development-System-JCIDS-18-Dec-2015.pdf

Collins, Bernie F., "Development of a Space Universal Modular Architecture (SUMO)," 2013 IEEE Aerospace Conference, Big Sky, Mont., March 2–9, 2013a.

———, "Space Universal Modular Architectures (SUMO): Commercial Space Industry's Dilemma," June 2013b.

Davis, Anthony, and Tom Ballenger, "Bridging the Valley of Death," *Defense AT&L Magazine*, January–February 2017, pp. 13–17.

Davis, Brian, "Interface Standards as an Enabler for Operational Responsive Space," presented at the Ground System Architectures Workshop, Redondo Beach, Calif., March 31–April 3, 2008. As of May 6, 2018: http://sunset.usc.edu/GSAW/gsaw2008/s2/davis.pdf

Divis, Dee Ann, "OCX Passes Deep Dive Review; GAO Says Program Risk Remains High," *Inside GNSS*, May/June 2017. As of August 1, 2017: http://www.insidegnss.com/node/5449

DoD Executive Agent for Space—*See* U.S. Department of Defense Executive Agent for Space.

Dubois, Thierry, "Eight Satellite Constellations Promising Internet Service From Space," *Aviation Week and Space Technology*, December 19, 2017. As of May 25, 2018: http://aviationweek.com/space/eight-satellite-constellations-promising-internet-service-space

Firesmith, Donald, "Open System Architectures: When and Where to Be Closed," *SEI Insights Blog*, Carnegie Mellon University, Software Engineering Institute, October 2015. As of September 7, 2017: https://insights.sei.cmu.edu/sei_blog/2015/10/open-system-architecture-when-and-where-to-be-closed.html

GAO—*See* U.S. Government Accountability Office.

Gustafson, Charles, and Siegfried W. Janson, *Think Big, Fly Small*, The Aerospace Corporation, September 1, 2014. As of September 7, 2017: http://www.aerospace.org/crosslinkmag/fall-2014/think-big-fly-small/

Hura, Myron, Gary McLeod, Lara Schmidt, Manuel Cohen, Mel Eisman, and Elliot Axelband, *Space Capabilities Development: Implications of Past and Current Efforts for Future Programs*, Santa Monica, Calif.: RAND Corporation, 2007, Not available to the general public.

Hyten, John, "Statement of John E. Hyten, Commander United States Strategic Command, Before the Senate Committee on Armed Services," April 4, 2017. As of September 2, 2017:
https://www.armed-services.senate.gov/imo/media/doc/Hyten_04-04-17.pdf

Insinna, Valerie, "US Air Force Creates Group to Recruit Cyber Nerds for Weapons Programs," *Defense News*, January 6, 2017a. As of July 14, 2017:
http://www.defensenews.com/articles/air-force-creates-new-group-to-recuit-cyber-nerds-for-weapons-programs

———, "Air Force Cancels Air Operations Center 10.2 Contract, Starts New Pathfinder Effort," *Defense News*, July 13, 2017b. As of July 15, 2017:
http://www.defensenews.com/articles/air-force-cancels-air-operations-center-102-contract-starts-new-pathfinder-effort

International Telecommunications Union, *Definition of "Open Standards,"* November 11, 2015. As of September 7, 2017:
http://www.itu.int/en/ITU-T/ipr/Pages/open.aspx

Joint Force Space Component Command Public Affairs, "CSPOC Established at Vandenberg AFB," July 24, 2018. As of October 11, 2018:
https://www.schriever.af.mil/News/Article-Display/Article/1582580/cspoc-established-at-vandenberg-afb/

Keithly, David M., and Stephen P. Ferris, "Auftragstaktik, or Directive Control, in Joint and Combined Operations," *Parameters*, Vol. 29, No. 3, Autumn 1999, pp. 118–133. As of October 14, 2018:
http://ssi.armywarcollege.edu/pubs/parameters/Articles/99autumn/keithly.htm

Kim, Yool, Elliot Axelband, Abby Doll, Mel Eisman, Myron Hura, Edward G. Keating, Martin C. Libicki, Bradley Martin, Michael E. McMahon, Jerry M. Sollinger, Erin York, Mark V. Arena, Irv Blickstein, and William Shelton, *Acquisition of Space Systems: Past Problems and Future Challenges, Volume 7*, Santa Monica, Calif.: RAND Corporation, MG-1171/7-OSD, 2015. As of August 23, 2018:
https://www.rand.org/pubs/monographs/MG1171z7.html

Lapham, Mary Ann, Suzanne Garcia-Miller, Lorraine Adams, Nanette Brown, Bart Hackemack, Charles (Bud) Hammons, Linda Levine, and Alfred Schenker, *Agile Methods: Selected DoD Management and Acquisition Concerns*, Pittsburgh, Pa.: Carnegie Mellon University, Software Engineering Institute, Technical Note CMU/SEI-2011-TN-002, October 2011.

Lapham, Mary Ann, Ray C. Williams, Charles Hammons, Daniel Burton, and Alfred Schenker, *Considerations for Using Agile in DoD Acquisition*, Pittsburgh, Pa.: Carnegie Mellon University, Software Engineering Institute, Technical Note, CMU/SEI-2010-TN-002, April 2010.

Larman, Craig, and Victor R. Basili, "Iterative and Incremental Development: A Brief History," *Computer*, Vol. 36, No. 6, IEEE Computer Society, June 2003, pp. 47–56. As of October 9, 2017:
http://ieeexplore.ieee.org/stamp/stamp.jsp?arnumber=1204375

Levine, Arnold, and Jeffrey Luck, *The New Management Paradigm: A Review of Principles and Practices*, Santa Monica, Calif.: RAND Corporation, MR-458-AF, 1994. As of August 23, 2018:
https://www.rand.org/pubs/monograph_reports/MR458.html

Lockheed Martin Corporation, *Lockheed Martin Demonstrates Continued Commitment to Air Force OMS Vision*, Palmdale, Calif., December 10, 2014. As of September 7, 2017:
http://www.lockheedmartin.com/us/news/press-releases/2014/december/lockheed-martin-demonstrates-commitment-to-air-force-oms-vision.html

Markuson, Nancy, and Joleen Flasher, *Adapting Agile Processes for Military Acquisition Programs*, briefing, McLean, Va.: MITRE, April 1, 2014.

McKinsey Agile Tribe, *The Five Trademarks of Agile Organizations*, New York: McKinsey and Company, December 2017.

Miller, Suzanne, Dan Ward, Mary Ann Lapham, Ray C. Williams, Charles Hammons, Daniel Burton, and Alfred Schenker, *Update 2016: Considerations for Using Agile in DoD Acquisition*, Pittsburgh, Pa.: Carnegie Mellon University, Software Engineering Institute, Technical Note CMU/SEI-2016-TN-001, December 2016.

MITRE, *Acquisition in the Digital Age: Analyze Requirements*, undated. As of September 7, 2017:
https://aida.mitre.org/models/agile-acquisition/msa-phase/analyze-requirements/

———, *Systems Engineering Guide*, Bedford, Mass., 2014.

Modigliani, Peter, and Su Chang, *Defense Agile Acquisition Guide: Tailoring DoD IT Acquisition Program Structures and Processes to Rapidly Deliver Capabilities*, McLean, Va.: MITRE, March 2014.

Morgan, Tiffany, "Multi-Mission Satellite Operations Center Ground System Architecture (MMSOC GSA)," SMC/SDTC, February 29, 2012. As of May 6, 2018:
http://gsaw.org/wp-content/uploads/2013/07/2012s11a_morgan.pdf

NASA, *Goddard Tech Transfer News*, Vol. 9, No. 2, 2011, pp. 4–5. As of May 6, 2018:
https://gmsec.gsfc.nasa.gov/images2/TechTransferNews-Spring11.pdf

National Defense Authorization Act for Fiscal Year 2018, Pub. L. No. 115-91, § 1601, December 12, 2017, 131 Stat. 1720.

Northern, Carlton, Kathleen Mayfield, Robert Benito, and Michelle Casagni, *Handbook for Implementing Agile in Department of Defense Information Technology Acquisition*, McLean, Va.: MITRE, Technical Report MTR100489, December 15, 2010.

Oar, Amber R., John P. Fitzsimmons, James P. Guthrie, Raymond A. Hoffman, Andrew J. Metzger, Carl J. Nelson, Taylor J. Olson, and Matthew J. Postupack, "DoD Acquisitions Reform: Embracing and Implementing Agile USAF," *Air and Space Power Journal*, Vol. 29, No. 6, 2015. As of August 23, 2018: https://www.airuniversity.af.mil/ASPJ/Display/Article/1152264/volume-29-issue-6-nov-dec-2015/

Oberndorf, Tricia, Mary Ann Lapham, Michael Bandor, and Charles "Bud" Hammons, "Finding Discipline in an Agile Acquisition Process," briefing presented at the 23rd Systems and Software Technology Conference (SSTC), May 16–19, 2011, Salt Lake City, Utah, Carnegie Mellon University, Software Engineering Institute, 2011.

Office of Defense Procurement and Acquisition Policy, *Guidance on Using Incentive and Other Contract Types*, memorandum, Washington, D.C., April 1, 2016. As of September 7, 2017: http://www.acq.osd.mil/dpap/policy/policyvault/USA001270-16-DPAP.pdf

Office of the Assistant Secretary of the Air Force for Acquisition, Technology, and Logistics, Science, Technology, and Engineering Directorate staff, discussion on prototyping and experimentation, June 23, 2017.

Office of the Deputy Assistant Secretary of Defense-Systems Engineering, "Initiative: Modular Open Systems Approach," webpage, undated. As of April 18, 2017: https://www.acq.osd.mil/se/initiatives/init_mosa.html

Operationally Responsive Space Office staff, teleconference discussion on ORS lessons on responsive acquisition, March 9, 2017.

Orbital ATK staff, discussion on MOSA, agile acquisition, and prototyping, Arlington, Va., June 27, 2017.

OTI—*See* Air Force Office of Transformational Innovation.

Parry, Daniel, "NRL at Blossom Point Celebrates 60 Years of Service," *The U.S. Naval Research Laboratory: News Releases*, June 14, 2016. As of May 6, 2018: https://www.nrl.navy.mil/news/releases/nrl-blossom-point-celebrates-60-years-service

Raymond, John W., "Fiscal Year 2018 Priorities and Posture of the National Security Space Enterprise," Department of the Air Force Presentation to the Subcommittee on Strategic Forces House Armed Services Committee, U.S. House of Representatives, May 19, 2017. As of September 2, 2017: http://docs.house.gov/meetings/AS/AS29/20170519/105974/HHRG-115-AS29-Wstate-RaymondJ-20170519.pdf

Reagan, Rex B., and David F. Rico, "Lean and Agile Acquisition and Systems Engineering: A Paradigm Whose Time Has Come," *Defense AT&L*, November–December 2010, pp. 49–52.

SAF/AQR—*See* Office of the Assistant Secretary of the Air Force for Acquisition, Technology, and Logistics, Science, Technology, and Engineering Directorate.

Sledge, Carol, "A Discussion on Open Systems Architecture," *SEI Insights Blog*, Carnegie Mellon University, Software Engineering Institute, November 23, 2015. As of September 7, 2017: https://insights.sei.cmu.edu/sei_blog/2015/11/a-discussion-on-open-systems-architecture.html

SMC—*See* Space and Missile Systems Center.

SMC/AD—*See* Space and Missile Systems Center, Advanced Systems and Development Directorate.

SMC/GP—*See* Space and Missile Systems Center, Global Positioning Systems Directorate.

SMC/PI—*See* Space and Missile System Center, Program Management and Integration Directorate.

SMC/RS—*See* Space and Missile Systems Center, Remote Sensing Systems Directorate.

Space and Missile Systems Center, "Maturation of Critical Technologies for Space Battle Management Command and Control (BMC2)," *Broad Agency Announcement (BAA)*, FA8819-17-BMC2-BAA, May 22, 2017.

Space and Missile Systems Center, Advanced Systems and Development Directorate, "Enterprise Ground Services," briefing, undated.

Space and Missile Systems Center, Advanced Systems and Development Directorate staff, discussion on challenges and ways to improve acquisition to support the SEV, El Segundo, Calif., January 20, 2017.

Space and Missile Systems Center, Global Positioning Systems Directorate staff, discussion on challenges and ways to improve acquisition to support the SEV, El Segundo, Calif., February 8, 2017.

Space and Missile Systems Center, Launch Enterprise Directorate staff, discussion on prototyping and Other Transaction Authority, El Segundo, Calif., April 18, 2017.

Space and Missile Systems Center, Program Management and Integration Directorate staff, discussion on challenges and ways to improve acquisition to support the SEV, El Segundo, Calif., February 8, 2017.

Space and Missile Systems Center, Remote Sensing Systems Directorate staff, discussion on challenges and ways to improve acquisition to support the SEV, El Segundo, Calif., February 7, 2017.

Space and Missile Systems Center, Space Defense Task Force, "Space Defense Task Force All Call," briefing, May 2017.

Space and Missile Systems Center, Space Defense Task Force staff, discussion on ESBMC2, El Segundo, Calif., June 30, 2017.

Space and Missile Systems Center, Space Superiority Systems Directorate staff, discussion on lessons from JMS Increment 2, El Segundo, Calif., January 20, 2017.

Stecher, Brian M., Frank Camm, Cheryl Damberg, Laura Hamilton, Kathleen Mullen, Christopher Nelson, Paul Sorenson, Martin Wachs, Allison Yoh, and Gail Zellman, *Toward a Culture of Consequences: Performance-Based Accountability Systems for Public Services*, Santa Monica, Calif.: RAND Corporation, MG-1019-EDU, 2010. As of October 24, 2018:
https://www.rand.org/pubs/monographs/MG1019.html

Troup, Tim, "Acquisition Truths from the Trenches," *Defense AT&L*, May–June 2010, pp. 42–45.

U.S. Air Force, "Enterprise Space Battle Management, Command and Control System (Joint Space Operations Center Mission System Increment 3)," Report to the Congress Defense Committees, April 2017.

U.S. Department of Defense, *Program Manager's Guide to Open Systems*, Version 2.0, Washington, D.C., September 2004. As of October 4, 2018:
http://www.acqnotes.com/Attachments/Program%20Managers%20Guide%20to%20Open%20Systems,%20Sept%202004.pdf

———, *Directive 5000.01: The Defense Acquisition System*, Washington, D.C., November 20, 2007. As of October 4, 2018:
https://www.dau.mil/guidebooks/Shared%20Documents%20HTML/DoDD%205000.01.aspx

———, "New Joint Interagency Combined Space Operations Center to Be Established," press release, Release No: NR-352-15, September 11, 2015. As of September 7, 2017:
https://www.defense.gov/News/News-Releases/News-Release-View/Article/616969/new-joint-interagency-combined-space-operations-center-to-be-established/

———, *Operation of the Defense Acquisition System*, Washington, D.C.: Department of Defense Instruction 5000.02, Change 2, February 2, 2017. As of October 4, 2018:
http://www.acqnotes.com/wp-content/uploads/2014/09/DoD-Instruction-5000.02-The-Defense-Acquisition-System-2-Feb-17-Change-2.pdf

———, Defense Standardization Program (DSP) Procedures, DOD 4120.24-M, December, 2017. As of October 4, 2018:
http://www.esd.whs.mil/Portals/54/Documents/DD/issuances/dodm/412024m.pdf

U.S. Department of Defense Executive Agent for Space, "Implementation Plan for Operationally Responsive Space," April 2008, Not available to the general public.

U.S. Government Accountability Office, *Defense Technology Development: Management Processes Can Be Strengthened for New Technology Transition Programs*, Washington, D.C., GAO-05-480, June 17, 2005.

———, *Best Practices: Stronger Practices Needed to Improve DOD Technology Transition Processes*, Washington, D.C., GAO-06-883, September 14, 2006.

———, *Software Development: Effective Practices and Federal Challenges in Applying Agile Methods*, Washington, D.C.:, GAO-12-681, July 2012.

———, *Defense Technology Development: Technology Transition Programs Support Military Users, but Opportunities Exist to Improve Measurement of Outcomes*, Washington, D.C., GAO-13-286, March 7, 2013.

———, *DOD Rapid Innovation Program: Some Technologies Have Transitioned to Military Users, but Steps Can Be Taken to Improve Program Metrics and Outcomes*, Washington, D.C., GAO-15-421, May 7, 2015a.

———, *Defense Advanced Research Projects Agency: Key Factors Drive Transition of Technologies, but Better Training and Data Disseminations Can Increase Success*, Washington, D.C., GAO-16-5, November 18, 2015b.

———, *Weapon Systems: Prototyping Has Benefited Acquisition Programs, but More Can Be Done to Support Innovation Initiatives*, Washington, D.C., GAO-17-309, June 27, 2017.

U.S. Navy, Office of Naval Research, "Elements of a Technology Transition Agreement (TTA)," undated. As of August 16, 2017:
http://www.onr.navy.mil/en/-/media/Files/03T/Elements-of-a-Technology-Transition-Agreement

———, *Future Naval Capabilities Guidebook 2017*, Arlington, Va., 2017. As of October 4, 2018:
https://my.nps.edu/documents/104914977/105848418/2017+Dist+A+ONR+Future+Naval+Capabilities+Guidebook+Mar+2017.pdf

Walden, Randall, "Air Force Rapid Capabilities Office Overview," Air Force Rapid Capabilities Office, June 16, 2016.

Ward, Dan, *FIRE: How Fast, Inexpensive, Restrained, and Elegant Methods Ignite Innovation*, New York: HarperBusiness, 2014.